Fiber Cable Testing, Certification, And Troubleshooting

- *Presentation Slides*

Revision 1.0

Eric R. Pearson, CFOS

Developed and Presented by

Pearson Technologies Inc.

4671 Hickory Bend Dr.

Acworth, GA 30102

www.ptnowire.com

770-490-9991

All rights reserved. Pearson Technologies Inc developed the following publication, Mastering Fiber Optic Network Design. All rights reserved. No part of this publication may be reproduced or distributed in any form or by any means, or stored in a database retrieval system without the prior written permission of the copyright holder. Copyright 2004, 2006, 2012, 2013, 2014 by Pearson Technologies Inc.

Disclaimer

All instructions contained herein are believed to produce the proper results when followed. However, these instructions are not guaranteed for all situations. In some cases, the author has expressed his opinions. Such opinions may not be technical fact.

Trademark Notice

All trademarks are the property of the trademark holder. Trademarks used in this document include Kevlar™ (Dupont), Hytrel™ (Dupont), Windows™ (Microsoft), QuickTime™ (Apple Computer), and ST-compatible (Lucent).

Fie: front pages/ Fiber Cable Testing, Certification, And Troubleshooting

Copyright Notice

© Pearson Technologies Inc., 2004, 2006, 2012, 2013, 2014

Fiber Cable Testing, Certification, and Troubleshooting — Introduction

Fiber Cable Testing, Certification And Troubleshooting

Developed And Delivered By
Eric R. Pearson, CFOS
FOA Master Instructor
BICSI Master Instructor
Pearson Technologies Inc.

Introduction

Six Goals

- Understand key concepts
- Recognize key terms
- Learn how to inspect, evaluate, and clean connectors
- Learn emergency connector replacement
- Learn how to test and measure key performance characteristics
- Learn how to interpret test values

Key Concepts

- Fiber characteristics that influence link performance
- Functions of cables
- Functions of connections

Connector

- Inspection
- Evaluation
- Cleaning
- Replacement

Key Tests And Measurements

- Power loss tests
  - Insertion loss
  - OTDR
- Reflectance
- Optical return loss

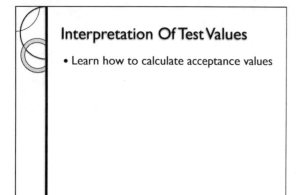
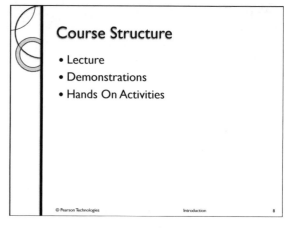
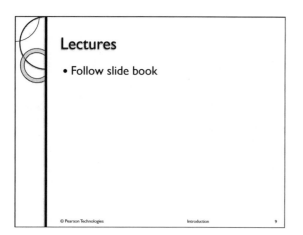
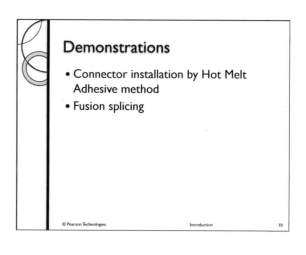

Fiber Cable Testing, Certification, and Troubleshooting	Introduction

Course Materials
- Slide book
- Professional Fiber Optic Installation
- Certification And Troubleshooting Fiber Optic Networks

11 Capabilities At Completion
- Perform insertion loss testing properly
- Interpret increased insertion loss properly
- Install no polish connectors
- Perform OTDR testing properly
- Identify features on an OTDR trace correctly
- Perform calculation of acceptance values properly

11 Capabilities At Completion
- Perform reflectance testing properly
- Perform ORL testing properly
- Perform connector evaluation
  ○ With microscope
  ○ With VFL
- Find concealed locations of high loss with VFL
- Troubleshoot and certify links quickly and properly

Schedule
- Days 1
  ○ 0800-1600 (approximately)
- Days 1-5
  ○ 0800-1530 (approximately)
- Breaks
  ○ 1015-1030
  ○ 1130-1230
  ○ 1415-1430
- Options
  ○ 0730 start/1630 finish
  ○ Short lunch for early finish

Your Duties
- Listen for understanding
  ○ Ask questions
  ○ Case study questions are wonderful
- Note package comprehensive
  ○ Minimal note taking necessary
  ○ Your slide text is my slide presentation
  ○ The text has almost all the words

End Of Introduction
Questions?
Comments?
Observations?

Fiber Cable Testing, Certification, And Troubleshooting   Chapter 1

Fiber Cable Testing, Certification And Troubleshooting
Developed And Delivered By
Eric R. Pearson, CFOS
FOA Master Instructor
BICSI Master Instructor
Pearson Technologies Inc.

Introduction To Fiber Networks
Chapter 1

The foundation for the rest of this program.

Objectives
- Learn basics
- Put fiber transmission in context

This Chapter
- Key concepts
- Advantages of fiber transmission
- Fiber networks
- Network components

Key Concepts
- Links create networks
- Links are duplex
- Networks are digital
- Link media

Links Create Networks
- Networks consist of a series of links
- Link is connection between network devices (i.e., network electronics)
- Devices convert signal to optical or to electrical
- All actions are on the link
- 'Link' is the most important term

© Pearson Technologies Inc.

Fiber Cable Testing, Certification, And Troubleshooting Chapter 1

Duplex Links

- All communication is duplex
- Duplex means
  - Two fibers per communication path or
  - Two wavelengths per path
- At link ends, most common cables are:
  - 2 simplex fiber patch cords
  - 1 duplex fiber path cord [recommended]
- Most common duplex patch cord is zip cord

Digital Networks

- All data, telephone, and CATV networks are digital
- Digital transmission enables unlimited regeneration without degradation of signal
- Language in this program is digital language
- Analog transmission is possible
- Analog transmission does not enable unlimited regeneration without degradation

Link Media

- While transmission over air, UTP, coax, via infra-red is possible
- Fiber transmission offers at least 8 advantages that motivate its use

Eight Advantages

- Nearly unlimited bandwidth
- Long transmission distance
- EMI/RFI immunity
- Low bit cost
- Dielectric construction
- Small size
- Light weight
- Easy installation

Fiber Networks

- Most data networks are mixed media

Mixed Media Networks

- Ethernet
- Fast Ethernet
- Gigabit Ethernet
- 10 Gigabit Ethernet
- 40 Gigabit Ethernet
- 100 Gigabit Ethernet
- Fiber Channel
- Asynchronous Transfer Mode (ATM)
- CATV networks

Four Current Versions Of Ethernet

- 1000BASE-SX, 1000BASE-LX
- 10GBASE-SX, 10GBASE-LX
- 40GBASE-SX, 40GBASE-LX
- 100GBASE-SX, 100GBASE-LX

Gigabit Ethernet

- Designations: 1000BASE-SX, 1000BASE-LX
- 1000 Mbps electrical rate
- 1250 Mbps optical rate- for accuracy
- Fibers allowed
  - 50μ
  - Singlemode
  - 62.5μ [not recommended]
- Transmission distance depends on fiber type and BWDP of multimode fiber
  - Range is 220-5000 m

10 Gigabit Ethernet

- Designations: 10GBASE-SX, 10GBASE-LX
- 10 Gbps electrical rate
- 12.5 Gbps optical rate- for accuracy
- Fibers allowed
  - 50μ
  - Singlemode
  - 62.5μ [not recommended]
- Transmission distance depends on fiber type and BWDP of multimode fiber
  - Range is 220-5000 m

40 And 100 Gbps Ethernets

- Designations
  - 40GBASE-SX, 40GBASE-LX, 100GBASE-SX, and 100GBASE-LX
- Multimode
  - 100 Gbps demultiplexed into 4 or 10, 10 Gbps signals
  - 10 Gbps multiplexed at receiver
  - 8 fibers for 40 Gbps
  - 20 fibers for 100Gbps
- Singlemode
  - Multiple wavelengths used (CWDM)
  - 2 fibers as in other duplex transmission schemes

Fiber Networks

- SONET and SDH
- FTTH
- FTTD
- PON
- FDDI

SONET

- SONET = Synchronous Optical Network
  - Is a North American standard
- Everywhere else:
  - Synchronous Digital Hierarchy (SDH)
- Data rates: multiples of 155.52 Mbps up to approximately 10 Gbps
- Rate designations: OC-x, x=1 to 192
  - OC-1= 155.52 Mbps
- Most common topology: ring

Fiber Cable Testing, Certification, And Troubleshooting — Chapter 1

SONET Ring

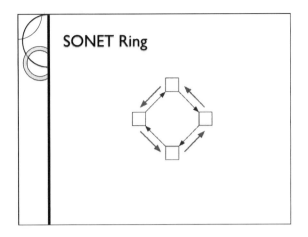

SONET Reliability

- Very high due to two paths around ring between all stations on ring
- Can have four paths around ring between all stations on ring
- Expensive technology
- 10 Gigabit Ethernet is replacing SONET
  - Advantage: significantly reduced cost

Three Types Of SONET Rings

- UPSR: Uni-directional Path Switched Ring
- BLSR: Bi-directional Path Switched Ring
- BLSR-4: Bi-directional Path Switched Ring With 4 Fibers

Common Features

- All stations on ring are add-drop multiplexors [ADM]
- Traffic travels on ring until it reaches its destination ADM
- All stations have same bandwidth
- Path determined by connections, not switches

UPSR SONET Ring

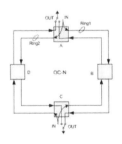

UPSR Operation

- Same traffic travels on both rings simultaneously
- Sum of bandwidth of all muxes is ≤ bandwidth of ring
- Consequence: can use half of available bandwidth
- In event of working path failure, traffic still available from protection path
- Lowest cost solution
- Lowest working bandwidth

UPSR With Break

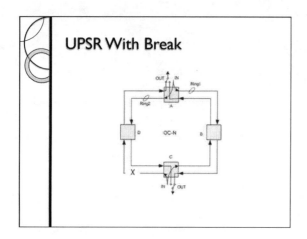

BLSR

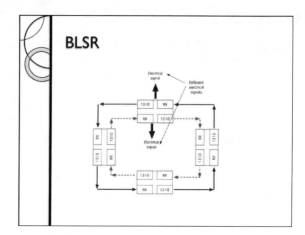

BLSR Operation

- Different traffic on both paths
- In event of failure of working path, traffic shifted to protection path
- Traffic on protection path dropped
- Increased bandwidth vs. UPSR

BLSR-4

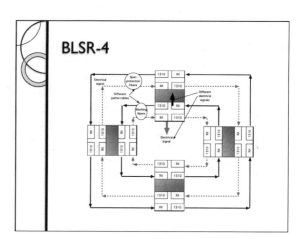

BSLR-4 Operation

- Have working and protection rings
- Increased fault tolerance
- Reduced recovery time
  - BLSR-2: ≤30 ms
  - BLSR-4: ≤10 ms
- Increased cost and complexity

Components

- Fiber
- Cable
- Connectors
- Splices
- Optoelectronics
- Transponders
- Hardware
- Passive devices

Fiber Cable Testing, Certification, And Troubleshooting — Chapter 1

Optoelectronic Designations

- GBIC (gigabit interface converter)
- SFP (small form-factor pluggable)
- Mini-GBIC
- Media converter
- NIC card
- XENPAK
- X2
- DOUBLE XENPACK

Transponders

- Some DWDM networks use transponders
- One optical signal is coupled directly into fiber
- All other optical signals are fed into transponders
- Transponders change input wavelength to narrow spectral width output wavelength

Transponder Example

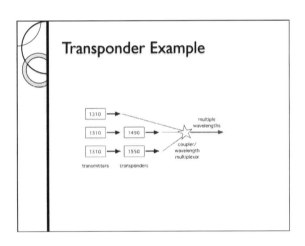

End Of Chapter

Questions?
Comments?
Observations?

Fiber Cable Testing, Certification, And Troubleshooting   Chapter 2- A Light Overview

Fiber Cable Testing, Certification And Troubleshooting
Developed And Delivered By
Eric R. Pearson, CFOS
FOA Master Instructor
BICSI Master Instructor
Pearson Technologies Inc.

A Light Overview
Chapter 2

Where Nothing Is Heavy
When Learning About Light!

This Chapter
- Learn the language of light
- Properties of light
- Behavior of light
- Numbers of light
- Significance of the different types of fiber (Chapter 3)

Basics
- Light exhibits specific behavior as it travels through fiber, connectors, and splices
- Light has properties

Behavior 1
- Rays
  - Travels in straight line
- Particles
  - Reflects/bounces
- Waves, or energy fields
  - Interference
  - Occupy a volume

Light Properties
- Wavelength
- Spectral width
- Speed
- Power
- Pulse width

© Pearson Technologies Inc.

Fiber Cable Testing, Certification, And Troubleshooting
Chapter 2 - A Light Overview

Wavelength
- When we think about light, we think of color
- The technical term for color is 'wavelength'
- Light has a periodic, or wave-like, nature
- The period of repetition is its wavelength
- Example
  - The distance from peak to peak, or from trough to trough, is the 'wavelength' of the waves in the water

Measuring Wavelength
- The wavelength of light, λ, is the measure of the distance between the peaks, or between the troughs, in nanometers (nm)
- Wavelength ranges from 780 nm to 1625 nm

Typical Wavelengths

Mode	Multimode nm	Multimode nm	Singlemode nm	Singlemode nm
Data	850 780	1300	1310	1550
CATV			1310	1550
Telephone			1310	1550
DWDM			1310-	1550
CWDM			1310	1490 1550 1625
WDM	850	1300	1310	1550
FTTH/PON			1310	1490 1550

Telephone Wavelength Bands

Band	Descriptions	Wavelength Range, nm
O	Original	1260-1360
E	Extended	1360-1460
S	Short	1460-1530
C	Conventional	1530-1565
L	Long	1562-1625
U	Ultra long	1622-1675

G.692 DWDM 100GHz Wavelengths

nm	nm	nm	
1528.77	1540.56	1552.52	Center wavelength
1529.55	1541.35	1553.33	
1530.33	1542.14	1554.13	
1531.12	1542.94	1554.94	
1531.90	1543.73	1555.75	
1532.68	1544.53	1556.55	
1533.47	1545.32	1557.36	
1534.25	1546.12	1558.17	
1534.04	1546.92	1558.98	
1535.82	1547.72	1559.79	
1536.61	1548.51	1560.61	
1537.40	1549.32	1561.42	
1538.19	1550.12	1562.23	
1538.98	1550.92	1563.05	
1539.77	1551.72	1563.86	

Significance of Wavelength
- The wavelength determines
  - Dispersion in the fiber determines
    - Signal accuracy
    - Maximum distance of transmission
  - Power loss in the fiber determines
    - Maximum distance of transmission

Fiber Cable Testing, Certification, And Troubleshooting   Chapter 2- A Light Overview

Wavelength And Dispersion
- With sufficient power at the receiver, if dispersion is excessive, the receiver will not convert the received power to an electrical signal that is identical to the input electrical signal
- More on this subject later

Wavelength And Power Loss
- If power loss is excessive, the receiver cannot convert the received power to an electrical signal that is identical to the input electrical signal

Light Properties
- Wavelength
- Spectral width
- Speed
- Power

Spectral Width
- When we use the term 'wavelength', we imply, incorrectly, that light has a single wavelength
- In all communications systems, the opposite state is true
  ◦ Light includes a range of wavelengths centered around a 'central', or 'peak', wavelength
- This range is the 'spectral width' of the light

Spectral Width
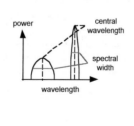

Significance
- Each wavelength in the spectral width travels at a different speed
- Spectral width is a specification for optoelectronics
  ◦ More on this later (see Dispersion)

© Pearson Technologies Inc.

Fiber Cable Testing, Certification, And Troubleshooting — Chapter 2- A Light Overview

Light Properties

- Wavelength
- Spectral width
- Speed
- Power
- Pulse width

Speed Of Light

- We all learned about the speed of light in school
  - The speed of light, c, is the speed at which light travels in a vacuum
- $c = 2.994 \times 10^8$ m/sec
- When light travels in any material, its speed drops

High Technologies..

- Use somewhat obscure terms for simple concepts
- Being a high technology, fiber optics will 'obscure' the term 'speed of light'
- For speed of light, we use
  - Index of refraction (IR) or
  - Refractive index (RI) or
  - η

Index Of Refraction

- Definition
  - RI= (speed of light in vacuum/speed of light in a material)
- With this definition, the RI is always greater than 1
- The RI of optical fibers ranges from approximately 1.46 to 1.52 (Appendix 1)
- Used to calibrate OTDR for accurate length and attenuation rate measurements

Sample IR Values

Product		850 nm	1300 nm	1310 nm	1550 nm
Draka MaxCap OM-2	50/125	1.482	1.477		
Draka MaxCap OM-3	50/125	1.482	1.477		
Draka MaxCap OM-4	50/125	1.482	1.477		
Laserwave 550-300	50/125	1.483	1.479		
LaserWave G+	50/125	1.483	1.479		
InfiniCor 50	50/125	1.481	1.476		
InfiniCor 62.5	62.5/125	1.496	1.491		
SMF-28e+				1.4676	1.4682
Draka singlemode G.652				1.4670	1.4680
OFS AllWave ZWP				1.4670	1.4680

IR Inaccuracy Examples

increment=	0.001				Length in m		
IR2	IR error	% error	100	500	1000	5000	10000
1.482			Error in distance to event, m				
1.483	0.001	0.0674309	0.07	0.34	0.67	3.37	6.74
1.484	0.002	0.1347709	0.13	0.67	1.35	6.74	13.48
1.485	0.003	0.2020202	0.20	1.01	2.02	10.10	20.20
1.486	0.004	0.2691790	0.27	1.35	2.69	13.46	26.92
1.487	0.005	0.3362475	0.34	1.68	3.36	16.81	33.62
1.488	0.006	0.4032258	0.40	2.02	4.03	20.16	40.32
1.489	0.007	0.4701142	0.47	2.35	4.70	23.51	47.01

© Pearson Technologies Inc.

Fiber Cable Testing, Certification, And Troubleshooting   Chapter 2- A Light Overview

Light Properties
- Wavelength
- Spectral width
- Speed
- Power
- Pulse width

Optical Power
- We can measure
  ◦ Power
  ◦ Power loss
- If we measure power, we measure absolute power
- If we measure power loss, we measure an relative power

Absolute Power
- Milliwatts
- dBm
  ◦ A power level relative to one milliwatt

dBm
- Definition:
  ◦ dBm= 10 log (power level/one milliwatt)
- Examples
  ◦ 0   dBm=   1 milliwatt
  ◦ 10 dBm= +10 milliwatts
  ◦ 20 dBm= +100 milliwatts
  ◦ -10 dBm=   0.1 milliwatts

dBm Uses
- dBm is used as measures of power
  ◦ Launched into a fiber by a transmitter
  ◦ Delivered to a receiver and
  ◦ Required for the receiver to function properly

Relative Power
- Definition
  ◦ dB= 10 log (power level/arbitrary power level)
- In testing, the 'arbitrary' power is a
  ◦ 'Reference' or
  ◦ 'Input' power level
- Relative power is measured in units of dB

© Pearson Technologies Inc.

Fiber Cable Testing, Certification, And Troubleshooting — Chapter 2- A Light Overview

dB Uses

- We use the term dB to indicate
  - Power loss of a component in a fiber link
    - Connectors specified ≤ 0.75 dB/pair
    - Attenuation rate specified ≤ 0.3-3.5 dB/km
  - Total loss in a fiber link
  - Maximum loss that can occur between a properly functioning transmitter and receiver
    - Transmitter-receiver pairs specified ≤ 12 dB

The Importance Of Power

- Power level drops as the pulse travels through fiber
- Power level at the input end of the link must be high enough that the receiver can convert the optical pulse to the same digital pulse at the transmitter, because…
  - As we all know, all electronic devices require a minimum power level to function properly

Volume

- In singlemode transmission, waves carry energy in the form of a field
- Energy fields exist within a volume

Light Properties

- Wavelength
- Spectral width
- Speed
- Power
- Pulse width

Pulse Width

- Each pulse has width
- Width results from time to turn on and off a transmitter
- This width must be less than that of the time interval allowed by the data rate
- In OTDR testing, pulse width determines length of fiber that can be tested

Light Behavior

- Reflection
- Refraction
- Dispersion
- Attenuation
- Skew

Reflection-Two Types

- We all recognize reflections
- For example, we see the sky reflected in a lake, a total reflection
- We see ourselves dimly reflected when we look through a closed window, a partial reflection
- Both of these types of reflections explain the behavior of light in optical fibers and their connections

Total Reflection

- The sky is reflected in a lake because there is a change in the speed of light/IR at the air-water boundary
- This change results in reflection as long as the angle of reflection is proper
- In the terms of physics, we see the sky as long as the rays of light strike the water within a critical angle
- If we look closer and closer to our feet, at some increased angle, which we call the 'critical angle', we stop seeing the sky and see into the water

The Critical Angle

- The maximum angle to the surface of the water at which we can see the sky is the critical angle
- Multimode fibers confine light because of a critical angle
- Critical angle is a characteristic and specification of multimode fibers

Critical Angle Reflection

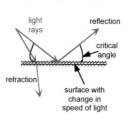

High Technologies..

- Use somewhat obscure terms for simple concepts
- For the term 'critical angle', we use
  - Numerical aperture or NA
- The NA, a measure of the critical angle, is defined
  - NA= sine (critical angle)
- Since the sine of an angle, in degrees, is a dimensionless number
  - NA is dimensionless
  - One of two dimensionless numbers in fiber optics

NAs

Critical Angle, °	NA
8.05	0.140
11.53	0.200
13.96	0.275

Partial Reflection

- Our reflection in the surface of a closed window is a Fresnel reflection
- A Fresnel reflection is a partial reflection
- A Fresnel reflection occurs any time light moves from one material, with an RI, to another material, with a different RI
- Connectors and splices create locations at which the RI can change
  - Resulting in reflections
- Such partial reflections can result in signal transmission errors

Fresnel Reflections In Connectors

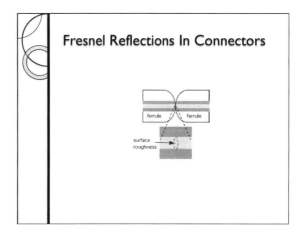

Significance

- Excessive reflection can result in transmission errors

Light Behavior

- Reflection
- Refraction
- Dispersion
- Attenuation
- Skew

Refraction

- Refraction, or bending, of light is common
- A pencil in a glass of water appears to be bent
- This bending occurs whenever light moves from a material, with one RI, to a material with a different RI
- The rule: as light moves from a higher RI to a lower RI material, it refracts towards the higher RI material

Example Of Refraction

Significance Of Refraction

- This bending explains the reduced dispersion that occurs in graded index multimode fibers (Chapter 3)

Light Behavior

- Reflection
- Refraction
- Dispersion
- Attenuation
- Skew

Five Forms Of Dispersion

- Modal dispersion
- Chromatic dispersion (CD)
- Material dispersion
- Waveguide dispersion
- Polarization Mode Dispersion (PMD)

Modal Dispersion

- Largest of five types
- Is path dispersion
- Occurs when different rays of light arrive at the fiber end at different times due to different travel paths
- This dispersion occurs in multimode fibers, but not in singlemode fibers

Path Dispersion

- All light rays in a pulse enter the fiber at the same time
- Options
  - All rays can travel the same path length
  - Rays can travel multiple paths

Same Path Means Same Path Length

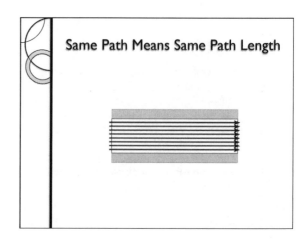

Fiber Cable Testing, Certification, And Troubleshooting   Chapter 2- A Light Overview

Same Path Length

- Results in 'single path' transmission
- We call this singlemode transmission
- Also known as 'monomode' transmission

Multiple Path Lengths

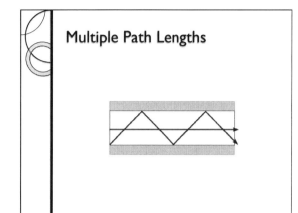

Multiple Path Lengths

- Results in multi path transmission
- We call this multimode transmission
- Significance
  ◦ Rays arrive at output end of fiber at different times
  ◦ This cause of dispersion is modal dispersion
    · The word 'mode' can be crudely translated to 'path'

Number Of Paths Determined

- By the fiber (Chapter 3)
- Increasing number of path lengths results in increasing dispersion

Light Behavior

- Reflection
- Refraction
- Dispersion
- Attenuation
- Skew

Attenuation

- Is power loss
- All network components have power loss
- In fiber, we refer to power loss as attenuation rate (Ch. 3)
- In connectors and splices, we refer to loss as loss

Fiber Cable Testing, Certification, And Troubleshooting   Chapter 2- A Light Overview

Light Behavior
- Reflection
- Refraction
- Dispersion
- Attenuation
- Skew

Skew
- The index of refraction is a nominal value, not an exact value
- Two optical signals on separate fibers will have slightly different IRs and travel times
- The difference in travel times between fibers is optical 'skew'
- Skew is a specification for multimode cables used for 40 and 100 G Ethernet transmission

Summary
- Light has characteristics, with performance values and units of measure

Characteristics, Terms, Units
- Wavelength and spectral width with units of nanometers (nm)
- Power, with units of dBm or dB
- Speed of light in an optical material with the dimensionless index of refraction and
- Critical angle with the dimensionless numerical aperture

End Of Chapter 2

Questions?
Comments?
Observations?

© Pearson Technologies Inc.

Fiber Cable Testing, Certification And Troubleshooting

Developed And Delivered By
Eric R. Pearson, CFOS
FOA Master Instructor
BICSI Master Instructor
Pearson Technologies Inc.

Fiber
Chapter 3

High Fiber Means High Throughput!

This Chapter

- Learn to fiber specifications
- Learn options
- Learn typical values
- Learn language

Fiber Basics

- Structure
- Types and Characteristics
- Performance
- Designations
- Bend Insensitive Fibers
- Fiber Specifications

Function

- The fiber is the medium through which light travels
- The fiber's function is to guide the light between the transmitter and receiver with minimum signal distortion
- Minimum signal distortion means
  - Minimum power loss
  - Minimal signal dispersion

Three Regions In Structure

- The fiber provides its function through its structure
- The structure consists of at least two, but usually three, regions
  - Core
  - Cladding
  - Primary coating
- Key fact: core and cladding have different
  - Compositions
  - IRs

Fiber Structure

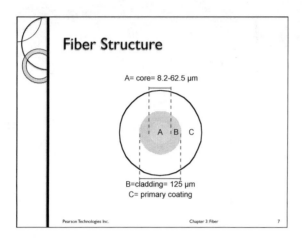

A = core = 8.2–62.5 µm
B = cladding = 125 µm
C = primary coating

Fiber

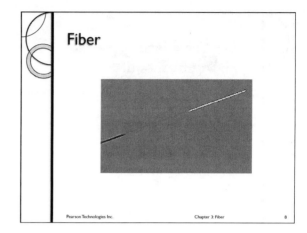

Functions of Regions

- The core is the central region of the fiber in which most of the light energy travels
- The cladding surrounds the core, confines the light to the core, and increases the fiber size so that it is easily handled
- The primary coating, formerly called the 'buffer coating', protects the cladding from mechanical and chemical attack so that the fiber retains its intrinsic high strength

Misconceptions

- Coating provides strength?
  - Allows fiber to retain its strength by protecting cladding from mechanical and chemical attack
- Cladding can be stripped?
  - Never with glass fibers

Three Primary Diameters

- The three regions of the fiber are characterized by their diameters, stated in microns (µ)
- Core diameter
  - The core diameter can range from 8.3µ to 62.5 µ
- Referred to as core diameter / cladding diameter
  - 50/125, 50/125/250, 125/50

Core Diameter Importance

- The core diameter is important to both the designer and installer
- The designer specifies the core diameter for two reasons
  - The core diameter must comply with that specified by the network standard
    - Core diameter determines transmission distance
  - The designer specifies the same core diameter throughout the network to avoid excess power loss at connections due to core diameter mismatch

Installer And Core Diameter

- The installer must know the core diameter in order to select the proper test leads for network testing
- If the installer uses test leads with a core diameter that does not match the core in the network, power loss measurements will be high
  - The installer will reject high loss links

Core Diameter History, Multimode

- Starting in 1985, standards for data networks, or short-haul LANs, required the use of the 62.5 μ fiber
- In 2010, the 50 μ fiber had a dominant market share because the most recent standards require it for its' superior transmission distance
- These standards are
  - Fiber Channel
  - 1000BASE-SX
  - 10GBASE-SX
  - 40 and 100 Gbps Ethernet

Recommendation

- Use 50μ
- No price advantage of 62.5μ
- Ignore 62.5μ unless significant advantage

Fourth Dimension

- NA
  - Is dimensionless, as you already know
- As a practical matter NA is related to core diameter
- When you match core diameter, you match NA

Core Diameter- NA Relationship

Core Diameter, μ	NA
8.2-10	0.14
50	0.20
62.5	0.275

Cladding Diameter

- The cladding is always 125μ
  - for our purposes

Primary Coating

- Primary coating diameter
  - Is usually 245 or 250μ
- As a practical matter, of little concern to designers or installers

Diameter Tolerances

- The core and cladding diameters are the most important fiber dimensions
- The designer has concern for four additional dimensions
  - Core diameter tolerance
  - Cladding diameter tolerance
  - Cladding non-circularity tolerance
  - Core offset tolerance

Diameter Tolerances

- The core and cladding diameter tolerances serve to limit the deviation of the actual diameters from nominal values
- Excessive deviation of these diameters results in
  - Excessive power loss
  - Increased installation cost

Core Diameter Deviations

- If the deviation of the core diameter from the nominal value is large, properly installed connections can exhibit excessively high power loss

Core Diameter Mismatch

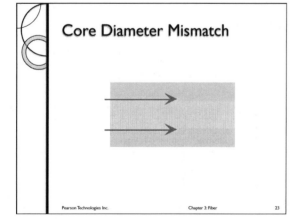

Cladding Diameter Tolerance

- Similarly, if the cladding diameter is small, the fiber cores may not be well aligned due to offset of fiber axes
- If the cladding diameter is large, the fiber may not fit into a connector

Fiber Cable Testing, Certification, And Troubleshooting — Chapter 3: Fiber

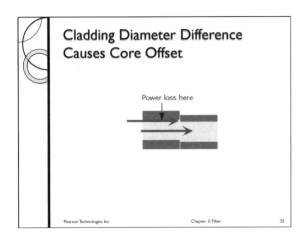

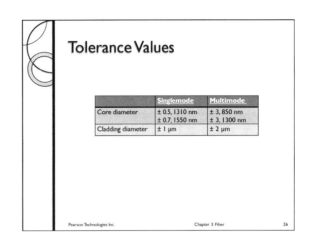

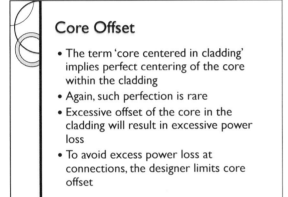

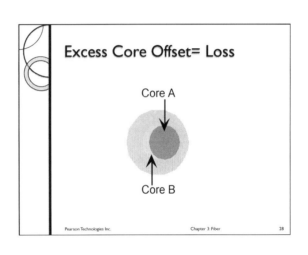

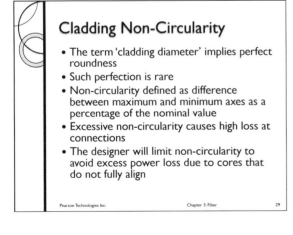

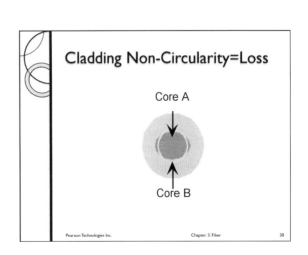

Fiber Cable Testing, Certification, And Troubleshooting — Chapter 3: Fiber

Tolerances

Fiber	Wavelength	Ovality	Offset, μm
Singlemode	1310 nm	≤ 1%	≤ 0.6
Singlemode	1550 nm	≤ 1%	≤ 0.6
Multimode		≤ 2%	≤ 3.0

Material Choices

- The core and cladding can be
  - Glass or plastic
- Fibers can have
  - Glass cores and glass cladding
  - Plastic cores and plastic cladding
  - Glass cores and plastic cladding
- Worldwide, most of the fibers have glass cores and glass cladding; i.e.: are 'all glass' (Our focus)

IMPORTANT FACT

- YOU CANNOT STRIP THE CLADDING!
- With cladding stripped, light escapes from side of fiber, not from end

Three Types

- Multimode step index
- Multimode graded index
- Singlemode
- Type depends on core diameter and structure of core

Multimode And Singlemode

- Multimode indicates that rays of light can travel multiple paths through the core
- Singlemode indicates that rays of light *behave as though* they are traveling along a single path through the core

Multimode Step Index Fibers

- The first type of optical fiber produced was a multimode plastic optical step index fiber (POF, SI)
- SI fibers have
  - A relatively large core
  - A single chemical composition in core
  - Limited bandwidth
- They are not allowed by any of the U.S. data standards

Step Index Composition

- Because the SI fiber has a single composition in the core, the speed of light is constant throughout the core
- Because the composition changes at the core-cladding boundary, rays of light can be reflected at that boundary

Step Index Core Profile (6-4)

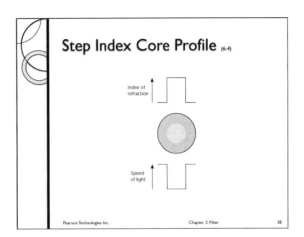

SI Reflection

- Rays of light that enter the end of the fiber at an angle less than or equal to the critical angle will be reflected at the cladding boundary
- Rays of light that enter the fiber at an angle greater than the critical angle pass through the boundary and into the cladding

Ray Reflection In SI Fiber

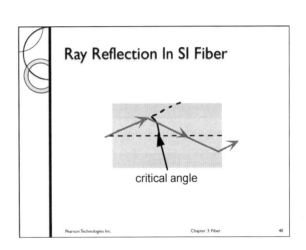

Total Internal Reflection

- Rays of light reflected from one side of the fiber travel to the opposite side
- At the opposite side, the rays are again reflected, as long as the rays remain within the critical angle defined by the NA of the fiber
- The name of the process is total internal reflection

Rays Outside NA Escape

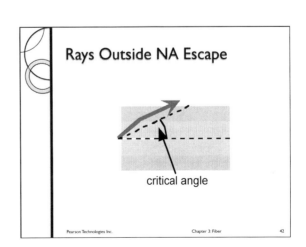

One Composition Consequence

- Rays of light in the same pulse can travel
  - Parallel to the axis of the fiber
  - Up to a maximum angle defined by the NA of the fiber
- These two rays enter the fiber at the same time, but they travel
  - Different paths and different path lengths
- Different rays arrive at the end of the fiber at different times: this is dispersion

Total Internal Reflection

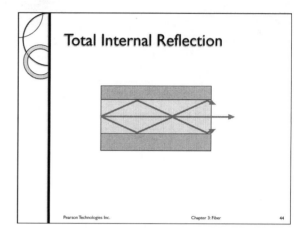

SI Dispersion Significant

- This difference in arrival time results in a significant amount of dispersion
- This dispersion results in a bandwidth that is too low to be of much use
- For this reason, the SI fibers are not specified for use in data networks

Multimode Graded Index

- Graded index (GI) core has
  - A relatively large core
  - Up to 2500 different chemical compositions
  - Compositions result in increased bandwidth

Capacity Measurement

- Bandwidth distance product (BWDP)
  - Units MHz-km
  - ≥ 500 MHz-km
- Alternative measure:
  - Differential modal delay (DMD)
  - Used for VCSELs
  - Laser optimized (LO) fibers
  - Equivalent to 1500-3700 MHz-km

GI Core Structure

- The GI fiber is constructed with up to 2500 compositions in the core
- These compositions are chosen so that the speed of light in the center of the core is lowest
  - Or, the IR is highest in the center of the core
- The speed of light in the core increases from the center to the cladding

GI Core Profile

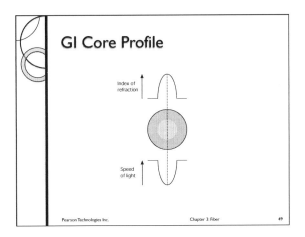

GI IR Profile
(Courtesy MIT Open Courseware)

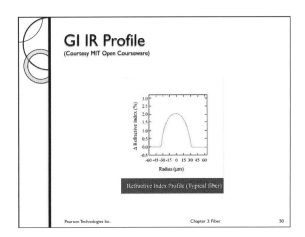

Dip In Profile

- Some early multimode fibers had an irregular profile
- This profile becomes important when laser transmitter sources used
- With such sources, a single pulse can split and bcome multiple pulses at receiver
- VCSELs, a form of laser, are designed to avoid such splitting [more on this later]
- OM-3, OM-4 fibers do not exhibit this dip

Profile Dip

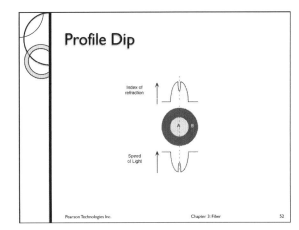

GI Compensation

- This core structure is compensates for difference in path lengths
- Axial rays at the center travel the shortest path length at the slowest speed
- Critical angle rays travel the longest path length at the highest average speed
- Can you see the compensation?

GI Core Construction

- This compensation allows rays that travel different path lengths to arrive at approximately the same time
- This compensation reduces the amount of dispersion
  - Bandwidth increases
- For this reason, GI multimode fibers are specified for use in data networks

'Curved' GI Paths

- A second consequence of this core structure is a 'bending' of the rays
- At each boundary between the layers, the rays bend/refract back towards the axis
  - Bending is from higher speed region to lower speed region
- Refraction reduces the difference in travel path distance of different rays
- In a GI fiber, light travels in paths that appear to be 'curved'

Ray Paths in GI Fiber

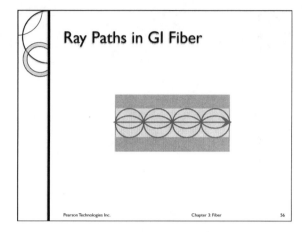

Multimode Designations (6-11)

- International fiber standard, ISO/IEC 11801, uses the designation 'OM', for optical fibers, multimode
- As a shorthand for fiber specifications
- Four OM-x designations approved
- Equivalent US standard: EIA/TIA 492xxxx
- Last designation approved
  - OM-4: EIA/TIA 492AAAD

Four Designations

OM	International	North America
OM-1	IEC 60793-2-10 Type A1b	TIA/EIA-492AAAA-A
OM-2	IEC 60793-2-10 Type A1a.1	TIA/EIA-492AAAB
OM-3	IEC 60793-2-10 Type A1a.2	TIA/EIA-492AAAC-A
OM-4	IEC 60793-2-10 Type A1a.3	TIA/EIA-492AAAD

OM Use

- OM-3 and OM-4 are laser optimized multimode (LOMM) fibers
- LO fibers enable multimode transmission to distances greater than those of standard multimode fibers
- Optimization is for multimode VCSELs in 1-100 Gbps Ethernet networks
- OM-3 and OM-4 have prices higher than those of OM-2 fibers
- This author recommends OM-3 and OM-4

OM Specifications

Designation	Dimensions	Bandwidth Distance Product, MHz-km		
		OFL		EMB
OM-x		850	1300	850
OM1	62.5/125	200	500	
OM2	50/125	500	500	
OM3	50/125	1500	500	2000
OM4	50/125	3500	500	4700

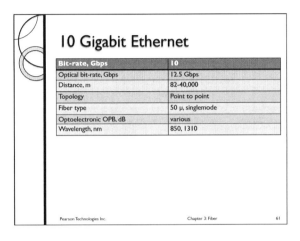
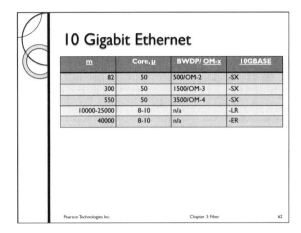
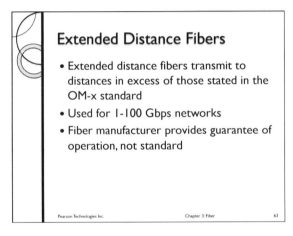
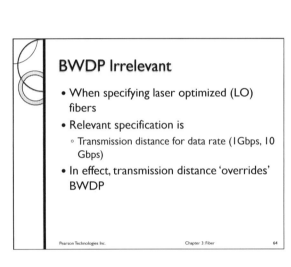
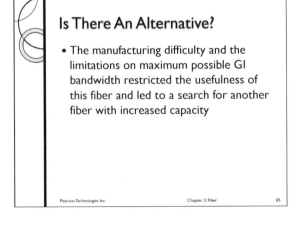

'Apparent' Singlemode Paths

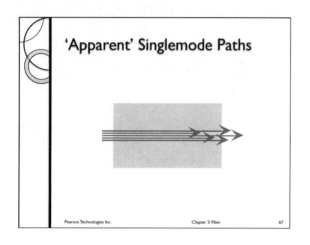

Singlemode Bandwidth

- Since all rays travel the same path, multiple path dispersion is eliminated
  - Resulting in *significantly* increased bandwidth
- Since such a fiber would allow a single path, this fiber became known as
  - Singlemode in North America
  - Monomode elsewhere
- Singlemode fiber is simpler, and less expensive, to make than is multimode fiber!

Singlemode Applications

- Singlemode fibers have
  - Essentially unlimited bandwidth
- This high capacity makes this fiber ideal for
  - Long haul, or inter-exchange telephone networks
  - CATV networks
  - FTTH networks

7 Singlemode Characteristics

- Long wavelength operation
  - 1310-1575 nm
- A 'cut off' wavelength
  - ≤ 1270 nm
- Small core diameter
  - 8.3-10 µ
- A 'mode field diameter' (MFD)
- Essentially unlimited bandwidth
- A 'dispersion rate' specification
- A zero dispersion wavelength

Mode Field Diameter

- Singlemode fibers are characterized by the mode field diameter (MFD)
- The MFD is the diameter within which most of the light energy travels
  - Is larger than the core diameter by approximately 1µ
  - Example: the Corning Inc. SMF-28 has a core diameter of 8.2 µ and a MFD of 9.2µ
- Some of the energy in travels in the cladding of a singlemode fiber

Core Diameter and MFD

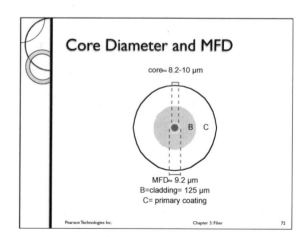

core≈ 8.2-10 µm
MFD≈ 9.2 µm
B=cladding= 125 µm
C= primary coating

Fiber Cable Testing, Certification, And Troubleshooting — Chapter 3: Fiber

MFDs

Fiber Type	MFD, µ
OFS Depressed Clad	8.8 ± 0.5 @ 1310 nm
OFS Matched Clad	9.3 ± 0.5 @ 1310 nm
OFS True Wave	8.4 ± 0.6 @ 1550 nm
Corning LEAF ®	9.2-10.0 @ 1550 nm
Corning SMF-28	9.2 ± 0.6 @ 1310 nm
Corning SMF-28	10.4 ± 0.8 @ 1550 nm
Corning MetroCor	8.1 ± 0.5 @ 1550

Significance Of MFD

- Mismatched mode field diameters can result in high power loss at connections
- Different singlemode fibers [G.652, G.653, G.655] have different mode field diameters

Light In Cladding?

- To understand this phenomenon, we step away from fiber optics for a minute.
- Imagine that you are working with a son or daughter on a high school science fair project. The project is to build a model of a single mode fiber.
- You and your son or daughter has decided to use a piece of pipe to simulate the fiber; the center to simulate the core; the wall, the cladding.
- You decide to use a ping-pong ball to model the photon traveling down the core of the fiber. You intend to shoot the ping-pong ball straight down the pipe, simulating singlemode transmission

Light In Cladding?

- However, light has some of the properties of an energy field. At this time, you have not modeled the properties of an energy field.
- After further consideration, you decide to rub the ping-pong ball with a piece of fur in order to create a static charge on the ball. The static charge creates an energy field.
- As the ball travels down the pipe, close to the inside wall, some of the static field travels in the wall. Analogously, some of the optical energy in a singlemode fiber travels in the cladding

Essentially Unlimited Bandwidth

- Essentially unlimited
  - ≤ 200 Tbps
- Source: Lucent Technologies, Approximately 2000

Singlemode Core History, 1

- First type, G.652
- Called: non-dispersion shifted
- Optimized for 1310 nm
- Can be used from 1310 nm -1550 nm
- 1-3 wavelengths
- Indicated by handset logo on cable jacket

G.652

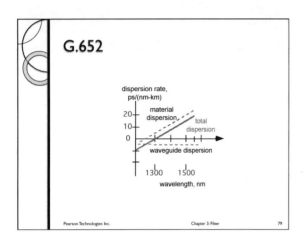

Likely Uses Of G.652

- G.652 within CO and cell tower site
- G.652 between CO and cell tower sites if you own fiber/cable
- Data networks (1-100 Gbps) use the G.652 fiber, which has a maximum bandwidth at 1310 nm

Singlemode Core History, 2

- Second type, G.653
  - Called: dispersion shifted (DS)
  - Optimized for 1550 nm
  - Can be used from 1310 nm -1550 nm
  - 1-8 wavelengths (CWDM)
- Not commonly used in US for telco transmission

G.653

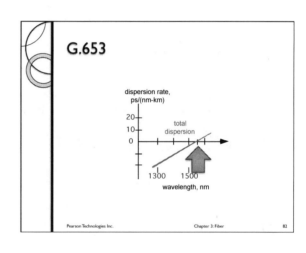

Singlemode Core History, 3

- Third type, G.655
- Called: Dispersion Shifted, Non-Zero Dispersion (DS-NZD)
- Optimized for approximately, but not exactly 1550 nm
- Optimized for >16 wavelengths (DWDM)
- Can be used from 1310 nm -1550 nm
- Two versions available
  - DS-NZD
  - Leaf™ (Large Effective Area Fiber, Corning Inc.)
    - DWDM fiber that allows increased power/wavelength

G.655 For DWDM

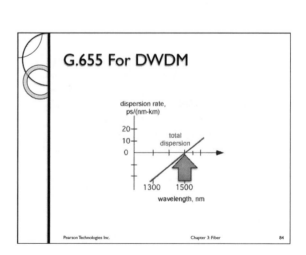

Fiber Cable Testing, Certification, And Troubleshooting — Chapter 3: Fiber

Likely Uses Of G.655

- DWDM links
- Has dispersion lower than that of G.652 at wavelengths near 1550 nm
- Has 1550 nm attenuation rate lower than that at 1310 nm [roughly 1/2]

If Singlemode Fiber Leased..

- Different fiber types have different MFDs and increased splice power losses
- Link with different fiber types will have dispersion and bandwidth that need
  - Measurements or
  - Calculations

MFDs

Fiber Type	MFD, μ
OFS Depressed Clad	8.8 ± 0.5 @ 1310 nm
OFS Matched Clad	9.3 ± 0.5 @ 1310 nm
OFS True Wave	8.4 ± 0.6 @ 1550 nm
Corning LEAF ®	9.2-10.0 @ 1550 nm
Corning SMF-28	9.2 ± 0.6 @ 1310 nm
Corning SMF-28	10.4 ± 0.8 @ 1550 nm
Corning MetroCor	8.1 ± 0.5 @ 1550

Must Match MFDs

- To avoid excess power loss at connections

Mismatch Power Loss

MFD, μ	power drop large to small dB	power drop large to previous dB
0.5 = increment		
8.0		
8.5	0.53	
9.0	1.02	0.50
9.5	1.49	0.47
10.0	1.94	0.45
10.5	2.36	0.42
11.0	2.77	0.40
11.5	3.15	0.39
12.0	3.52	0.37

Mismatched Fibers, Mismatched Dispersion Rates

- Link capacity or distance reduced by high dispersion in one of mismatched fibers

Mismatched Dispersions: G.652 And G.655

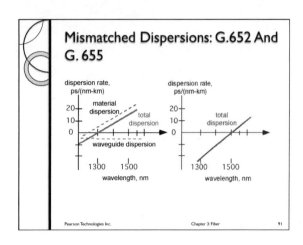

Performance

- As a pulse of light travels through a fiber, the pulse changes in two ways
  - The pulse width increases
  - The pulse height reduces
- These two changes give rise to the only two transmission characteristics of the fiber
  - Dispersion
  - Attenuation or power loss

Two Characteristics

- Dispersion
- Power loss

Dispersion

- We present dispersion with the assumption of digital transmission
- With the assumption of analog transmission, dispersion results in a smearing of the signal, making the signal less precise

Multimode Dispersion And Effective Bandwidth

- Six characteristics determine multimode dispersion
  - Fiber core diameter
  - NA
  - Wavelength of the light source
  - Spectral width of the source
  - Bandwidth distance product
  - Transmission distance

BWDP Subtlety

- BWDP determined under conditions of
  - Zero spectral width
  - Zero chromatic dispersion
- Reality: all optoelectronics have a non-zero spectral width and some chromatic dispersion

Consequence

- BWDP/distance ≠ effective bandwidth

Real Conversion Is Complex

- The equations take 1.5 pages to list!
- Better boot your computer!

Effective BW Example 1

- Link length: 1 km
- Fiber: 50/125
- 850 nm LED light source
- Spectral width: 30-50 nm
- 200 MHz-km BWDP = effective bandwidth, of approximately 83 MHz to 119 MHz
- BWDP/distance= 200 MHz?

Effective Bandwidth Vs. BWDP for 50/125 Fiber With 850 nm LED

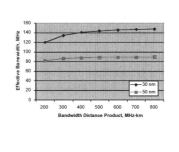

Two Conclusions

- Division of the BWDP/distance ≠ real bandwidth
  - Real bandwidth is much lower
- Little benefit from increased BWDP>200 MHz-km.
  - 800 MHz-km and spectral width of 50 nm results in effective BW= 89 MHz
  - A cable cost increase of 25-50 % results in increase in throughput of approximately 10 %
  - Not a good deal!

Effective BW Example 2

- Link length: 1 km
- Fiber: 62.5/125
- 850 nm LED light source
- Spectral width: 30-50 nm.
- 200 MHz-km BWDP = effective bandwidth of approximately 75-112 MHz

Effective Bandwidth for 62.5/125 Fiber, 850 nm LED

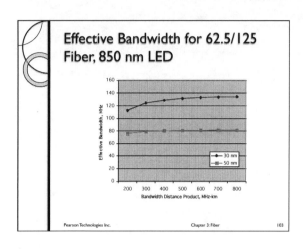

Conclusions 2

- Same as in Example 1
  - Effective bandwidth lower than expected
  - Little benefit from increased BWDP

Effective BW Example 3

- Link length: 1 km
- Fiber: 50/125
- 1300 nm LED light source
- Spectral width: 60-170 nm.
- 200 MHz-km BWDP = effective bandwidth, of approximately 180 MHz to 198 MHz.
- BWDP/distance= 200 MHz

Effective Bandwidth for 50/125 Fiber, 1300 nm LED

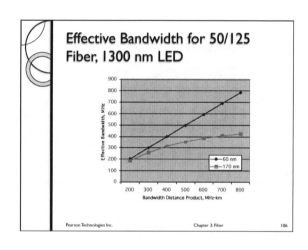

Conclusions 3

- Increased throughput at 1300 nm
- Effective bandwidth still below value from simple calculation
- Increased BWDP provides benefit
  - 800 MHz-km produces effective bandwidth of 420 MHz

Design Rule

- Use 850 nm optoelectronics if you can
  - 850 nm optoelectronics are less expensive than 1300 nm optoelectronics
- Use 1300 nm optoelectronics with multimode fiber only if you must

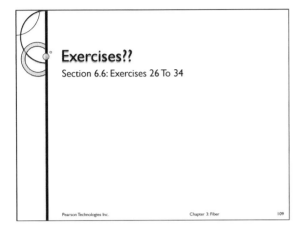

Exercises??
Section 6.6: Exercises 26 To 34

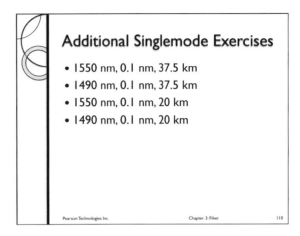

Additional Singlemode Exercises
- 1550 nm, 0.1 nm, 37.5 km
- 1490 nm, 0.1 nm, 37.5 km
- 1550 nm, 0.1 nm, 20 km
- 1490 nm, 0.1 nm, 20 km

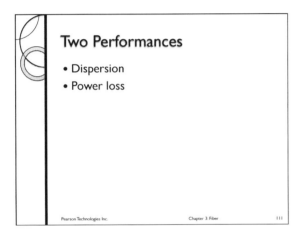

Two Performances
- Dispersion
- Power loss

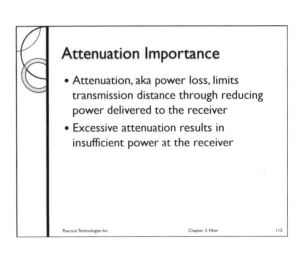

Attenuation Importance
- Attenuation, aka power loss, limits transmission distance through reducing power delivered to the receiver
- Excessive attenuation results in insufficient power at the receiver

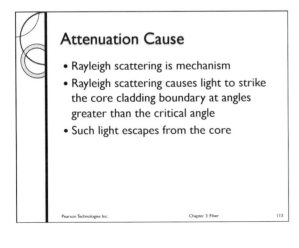

Attenuation Cause
- Rayleigh scattering is mechanism
- Rayleigh scattering causes light to strike the core cladding boundary at angles greater than the critical angle
- Such light escapes from the core

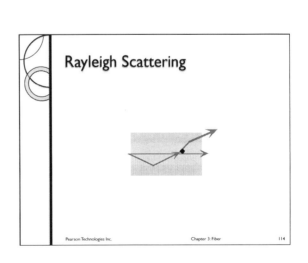

Rayleigh Scattering

Fiber Cable Testing, Certification, And Troubleshooting — Chapter 3: Fiber

Installer's Note

- Attenuation is the most important characteristic during installation, since installation mistakes can, and do, result in an increased power loss in the cable

Maximum And Typical

- The network designer uses the maximum attenuation rate as a cable specification
- The typical values are those that you should expect to see on properly designed, manufactured and installed cables
- The network designer uses both the maximum and typical attenuation rates to calculate the acceptance value, which is the power loss value which the installed cable must not exceed
- Goal is maximum reliability

Attenuation Rates, Maximum

Wavelength, nm	Core Diameter, μ	Attenuation Rate, dB/km
850	62.5	3.5
850	50	3.5
1300	62.5	1.5
1300	50	1.5
1310	8.2	0.5, 1.0
1550	8.2	0.5, 1.0

1310, 1550 nm Attenuation Rates

- 0.5 dB/km is for loose tube cables
- 1.0 dB/km is for tight tube cables

Attenuation Rates, Typical

Wavelength, nm	Core Diameter, μ	Attenuation Rate, dB/km
850	62.5	2.8-3.0
850	50	2.5-2.7
1300	62.5	0.7
1300	50	0.7
1310	8.2	0.30-0.35
1550	8.2	0.20

Note Attenuation Rates @ 1550 nm

- 1550 nm attenuation rates are lower than those at 1310 nm
- Standard does not reflect this advantage
- This reduced attenuation rate is one reason for using 1550 nm transmission

© Pearson Technologies Inc.

Fiber Cable Testing, Certification, And Troubleshooting — Chapter 3: Fiber

Singlemode Water Peak

- Attenuation rate is determined by wavelength
- In general, attenuation rate drops as wavelength increases
- Exception: there may be an increase in attenuation rate at approximately 1390 nm
- This increase is a water peak caused by water in the fiber

Zero Water Singlemode Fiber

- Water peak absent in zero water or no water fiber
- Lack of water peak an advantage in DWDM and CWDM applications

Attenuation Rate Vs. Wavelength
(Courtesy MIT Open Courseware)

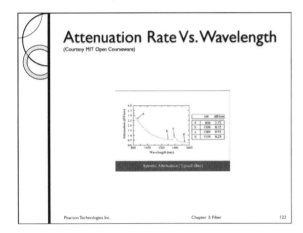

Singlemode Spectral Attenuation
(Courtesy Corning Inc.) (6-13)

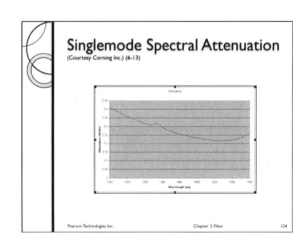

Singlemode Water Peak
(Courtesy Corning Inc.)

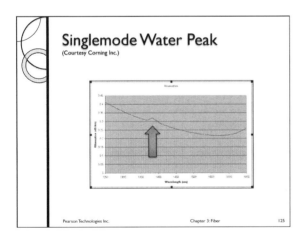

Attenuation Uniformity

- Attenuation rate non-uniformity specification consists of maximum loss at non-uniformity and number of non-uniformities allowed
- TIA/EIA-568-C provides values of loss
  - ≤ 0.1dB/point, singlemode
  - ≤ 0.2dB/point, multimode
- Number of non-uniformities is up to you
- Suggested number allowed: 1

Additional Fiber Types

- Bend insensitive [BI]
- Can be bend to radii smaller than standard fibers
- Have loss increases less than those of standard fibers
- Multimode and singlemode available

Bend Insensitive (BI) Fibers

- Singlemode BI fibers compatible with standard singlemode fibers
- Multimode BI fibers (BIMM) may not be compatible with standard fibers
- Loss testing method for multimode BI fibers may change in near future

Important Fact

- All numbers are qualified by a specific test procedure, called Fiber Optic Test Procedure
  - Aka FOTP-xxx or EIA/TIA-455-xxx

Examples Of Test Procedures

Mechanical & Environmental Characteristics:

♦ Crush Resistance	EIA-FOTP-41A	2200 N/cm	
♦ Impact Resistance	EIA-FOTP-25B	500 Impacts	
♦ Flexing	EIA-FOTP-104A	10000 Cycles	
♦ Cable Knot Test	EIA-FOTP-87B	Passed @ -40°C	
♦ Cable Twist Test	EIA-FOTP-85A	1000 Cycles	
♦ Fluid Immersion Test	EIA-FOTP-12A	JP-2 Fuel	
♦ Maximum Pulling Load	EIA-FOTP-33A		
	2 Fiber	1800 Newtons	
	4 Fiber	1800 Newtons	
♦ Maximum Safe Operating Load		600 Newtons	
♦ High and Low Temperature Bend Test			
	EIA-FOTP-37A Passed @ -40°C and +80°C		
♦ Minimum Bend Radius at Maximum Load		6 cm	
♦ Minimum Bend Radius Unloaded		3 cm	
♦ Polyurethane Jacket			
♦ Operating Temperature	EIA-FOTP-3	-40°C to +85°C	
♦ Storage Temperature (non-flexing)		-55°C to +85°C	
♦ Installation Temperature		-40°C to +85°C	
♦ Diameter (nominal)	4 Fiber 5.3 mm (0.210")	2 Fiber 5.3 mm (0.210")	
♦ Weight	42 lbs/km	39 lbs/km	
♦ Fiber Proof Test	100 KPSI		

Design Rules

- Use 850 nm fibers if you can
  - 850 nm optoelectronics are less expensive than 1300 nm optoelectronics
- Use 1300 nm optoelectronics with multimode fiber only if you must
- Use multimode fibers if you can
  - Multimode optoelectronics less expensive than singlemode optoelectronics
- Use singlemode fibers if you must

Design Rules- For Cell Towers

- Within tower sites, use OM-3 or OM-4
  - Reduced optoelectronic cost
  - No need to change fiber for future increase in bandwidth
  - Remember: for potential upgrade to 40 Gbps, install 8 fibers for each link
- Use singlemode fibers between tower sites
  - Reduced link cost
  - No need to upgrade in future

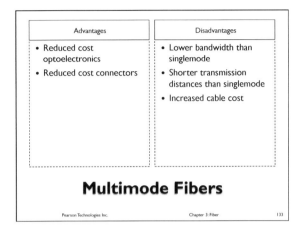

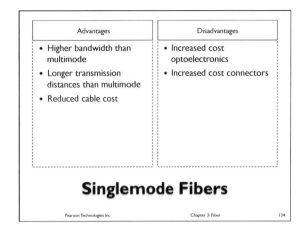

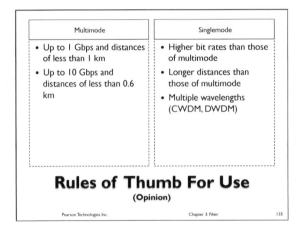

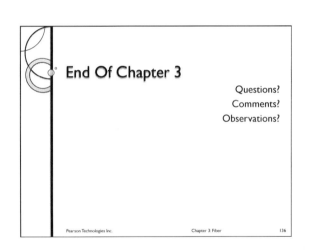

Fiber Cable Testing, Certification, And Troubleshooting   Chapter 4: Cables

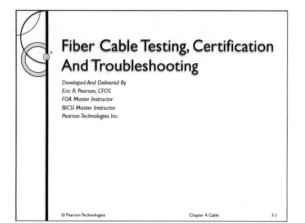

Fiber Cable Testing, Certification And Troubleshooting

Developed And Delivered By
Eric R. Pearson, CFOS
FOA Master Instructor
BICSI Master Instructor
Pearson Technologies Inc.

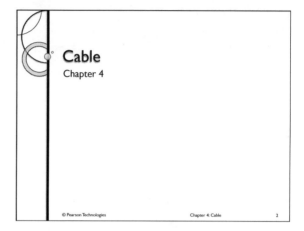

Cable
Chapter 4

This Chapter
- Learn
  - Performances
  - Types

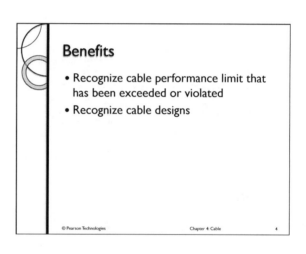

Benefits
- Recognize cable performance limit that has been exceeded or violated
- Recognize cable designs

Cable Basics
- Why cable the fiber?
- Imagine five miles of directly buried glass optical fiber with a diameter of 0.010"!
- While the fiber can easily perform the task of optical communication, it cannot, in and of itself, survive installation and use without additional protection
- The cable is the package that provides protection against damage

Three Protections
- Increased attenuation
- Breakage of fiber
- Delayed breakage of fiber
  - Delayed breakage is also known as delayed failure and static fatigue
  - Static fatigue is failure of glass materials under relatively low, long-term stress

© Pearson Technologies Inc.

Fiber Cable Testing, Certification, And Troubleshooting — Chapter 4: Cables

Fact To Remember

- Protection has limits
  - Conditions imposed on cable must not exceed rating of cable
  - Cable damage is related to exceeding cable rating

Protections

- Installation
- Environmental

Installation Characteristics

- Objective: enable installation without damage
- This is installer's concern
- Alternatively, this is a concern of operator when identifying link problems

Installation Requirements

- Installation load
- Short-term bend radius
- Outer jacket diameter
- Inner jacket diameter
- Buffer tube diameter
- Storage temperature range
- Installation temperature range

Load Damage

- Breakage of fiber
- Increased attenuation
- Delayed breakage of fiber
  - Delayed breakage is also known as delayed failure and static fatigue
  - Static fatigue is failure of glass materials under relatively low, long-term stress

Loads

- Short term, or installation, load
  - Short term: maximum load to be applied for short term; i.e., during installation
  - Range: 110-600 lbs-force (490-2700 N)
  - Important when pulling cable
    - Installer must limit load during pulling
  - Less important for installation in cable trays
- Long term, or use load
  - Long term: to be left on the cable for entire lifetime
  - Can be zero

Fiber Cable Testing, Certification, And Troubleshooting   Chapter 4: Cables

Bottom Line

- This value becomes a minimum requirement that cable must meet

Short Term Bend Radius

- Rule for short term bend radius ≥ 20 x cable diameter
- Can be critical in man holes
- TIA/EIA-568-C values
  - For inside cable
    - 2-4 fibers: ≥2" (50 mm at 50 lbs.-f/220 N)
    - ≥ 4 fibers: ≥20 x cable OD
  - For inside/outside and outside cable
    - ≥20 x cable OD
  - Use if appropriate

Cable Dimensions

- Outer jacket diameter
  - Important for cable in conduit or inner duct
- Inner jacket diameter must fit into the boot of the connector
  - One-fiber cables
  - Zip cord duplex
  - Breakout sub cables
- Buffer tube diameter
  - Must fit into the back shell of the connector

Environmental Protection

Cables must survive the environment within which they are installed.

Action Required

- If these conditions exist in environment in which cable is to be installed,
  - Specify performance level

Environmental Characteristics

- All Cables
- Indoor
- Outdoor

© Pearson Technologies Inc.

All Cables

- Operating temperature range
- Use load
- Vertical rise distance
- Long term bend radius
- Dielectric design

Temperature Range, Operating

- Definition: the range within which the cable will exhibit an attenuation rate below its maximum
- Additional benefit: materials will not deteriorate

Generic Performance

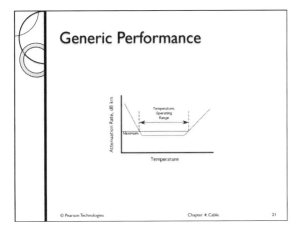

Two Examples

- Steam tunnel in hospital
- Concrete conduit system between buildings in Colorado

TIA/EIA-568-C Performance

- TIA/EIA-568-C references two cable standards
- Outside plant cables: ANSI/ICEA S-87-640-1999
- Inside plant cables: ANSI/ICEA S-83-596-2001
- Wider ranges exist for application specific cables, e.g., military and aircraft

TIA/EIA-568-C Temperature Range
(7-14)

Type	Range
Indoor	-20° C. to +70° C.
Plenum	0° C. to +70° C.
Outdoor	-40° C. to +70° C.

Use Load

- A use load specification is required if a load in applied to the cable by its environment
- Examples
  - Cables installed between widely spaced poles, buildings, power transmission towers, in risers
- Use loads range widely, depending on the cable type

Maximum Use Loads Available

Application	Values, pounds-force
1 fiber, raceway	23-35
1, fiber, conduit	67
6-12 fiber cable	33-330
Direct burial cable	132-180
Self-support cable	2000

Other Forms Of Use Load

- Cable span rating
- Wind speed rating
  - E.g., 150 mph hurricane rating
- Vertical rise distance

Maximum Vertical Rise Distances

Application	Values, feet
1 fiber, raceway	90
1, fiber, conduit	50-90
6-12 fiber cable	50-375
Heavy duty cable	1000-1640

Example

- 50 storey bldg at 12-15'/storey has a maximum vertical distance of 600-750 ft.

Long Term Bend Radius

- Definition: minimum radius to which cable will be bent for its lifetime while under no load
- Aka, use bend radius
- Always important
- Cable path imposes bend on cable
- Minimum conduit sweep or elbow radius may impose minimum bend radius on cable
- Size of enclosure used for storage loop imposes minimum bend radius on cable

Two Methods To Determine Radius

- Use values in TIA/EIA-568-C
  - Use if appropriate
- Determine actual value needed from knowledge of cable path

Use TIA/EIA-568-C

Type	Minimum Bend Radii
Indoors, 2-4 fibers	≥ 1"
Indoors, ≥ 4 fibers	≥10 x diameter
Outdoor and indoor/outdoor	≥10 x diameter

Bottom Line

- This value becomes a maximum requirement that cable must meet
- Wording: cable must have a minimum long term bend radius equal to or larger than this value

Dielectric Design

- Designers specify dielectric design frequently because of
  - Increased ease of installation
  - Increased safety
  - Reduced costs
    - Reduced costs result from elimination of installation and maintenance of grounds and bonds

Indoor Cable Requirements

- NEC rating
  - Not applicable to cell phone tower installations
- Zero halogen or halogen free

Two Classes: OFN And OFC

- OFNx: a dielectric structure
- OFCx: a conductive structure

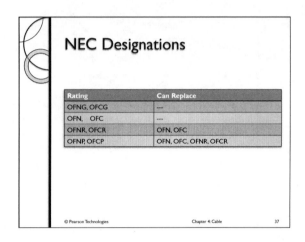
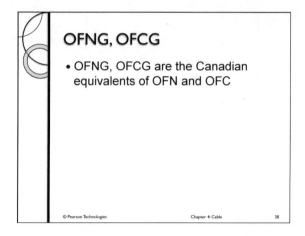
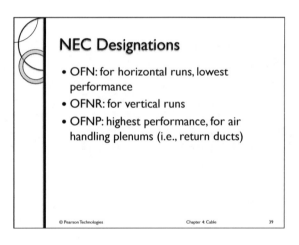
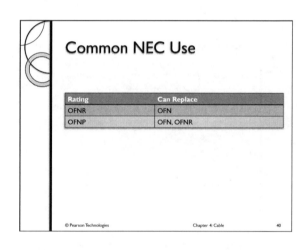
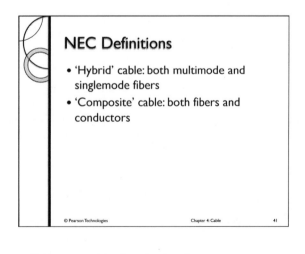
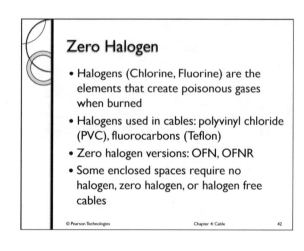

Enclosed Spaces

- Data centers
- Tunnels
- Aircraft
- Ships
- Submarines
- Mobile platforms
- Tanks

Cell Phone Systems

- Zero halogen may be best choice for CO

Outdoor Cable Protections

Moisture	Rodent	UV
High voltage	Lightning	Impact
Crush, long term	Crush-short term	Radiation
Hydrogen	Vibration	Flame
Gas flow	Filling flow	Abrasion
Flexing	Steam	Chemical

Moisture Resistance: 3 Problems

- Cables channeling water into the electronics
- Fiber breakage from water freezing inside cable
- Degradation of fiber strength from exposure of cladding to (non-neutral) ground water

Two Solutions

- Water blocking gels and greases are placed inside of buffer tubes and in all unfilled spaces outside the buffer tubes
  - Considered undesirable
  - Gel required in REA-financed systems
- Super-absorbent polymer (SAP) tapes and yarns are placed inside the buffer tubes and under the jacket
  - No cost increase
  - Reduces installation labor cost

Filled And Blocked

- 'Gel filled and grease blocked'
  - Gel filling inside buffer tubes
  - Grease in all other empty areas
- Only method used from mid 1970's-1994
- Disadvantage: significant increase in cost for end preparation of cables

SAP

- Dominant method
- Designations: 'gel free, grease free', or 'dry water blocked'
- SAP converts water into gel
  - 'Water super absorbents with the ability to absorb up to 900g water by one gram material have been prepared' (http://www.springerlink.com/content/lvyx6xqyaaujhjuc/)
- Gel swells into jacket cracks, preventing further moisture ingress

SAPs
(Courtesy Corning Cable Systems Inc.)

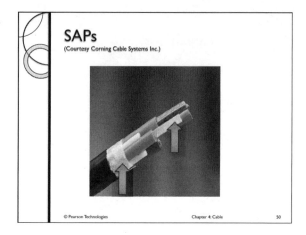

Bottom Line (Opinion)

- If moisture resistance required, specify dry water blocked

Rodent Resistance
(Courtesy Lucent Technologies)

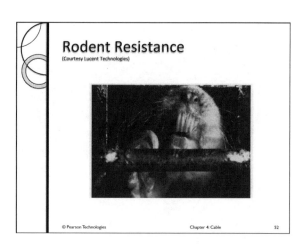

Two Alternatives

- Armored cable
- Install non-armored cable in inner duct or conduit

Armor

- Advantages
  - High crush and rodent resistance
  - Can locate underground cable
- Disadvantages
  - Increased cost
  - Reduced flexibility
  - Increased installation and maintenance costs:
    - Must ground and bond

Fiber Cable Testing, Certification, And Troubleshooting   Chapter 4: Cables

Armored Option 1: No Internal Jacket

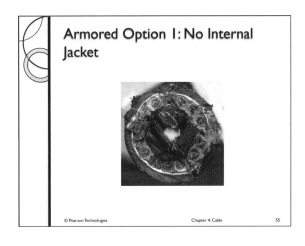

Armored Option 2: With Internal Jacket

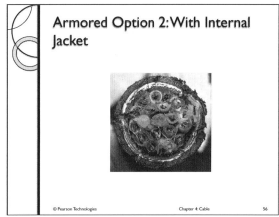

Alternative: Install In Inner Duct

- Inner duct at least 1.5" in diameter
- Duct is too large for rodent to chew

Inner Duct

- Advantages
  - Dielectric cable path
  - No grounding or bonding
- Disadvantages
  - Must provide location method (wire)
  - Potential increased installation cost: two installation steps

Recent Addition

- Dielectric armor

Note

- If time is sufficient and cable handling equipment can handle inner duct, you can order the inner duct extruded over the cable
- Advantage: a single installation step

Fiber Cable Testing, Certification, And Troubleshooting   Chapter 4: Cables

Bottom Line

- If dielectric cable path is required, use inner duct and non armored cable
- If total installed cost favors armored cable, specify stainless steel armor
- If total installed cost favors inner duct, specify inner duct and non armored cable

UV Resistance

- Ultraviolet light causes degradation of many polymers
- Outdoor cables need to be resistant to such degradation

Typical Solution

- High density black polyethylene (HDPE)
- Benefits:
  - Low cost
  - Long life
    - HDPE resists most conditions in outdoor environments
- Disadvantage: HDPE does not meet the requirements of the National Electric Code (NEC)
  - It's a member of the paraffin family!

Additional Solution

- Indoor-outdoor cables with single jacket that provides
  - UV resistance
  - Compliance with the NEC

Indoor-Outdoor Cable

- Advantage: elimination of indoor outdoor cable connection
- Beneficial in networks in which outdoor cable must feed optoelectronics more than 50' from entrance point
- Estimated cost savings: $20-$40/fiber/end
- Disadvantage: increased cable cost

Example

- Increased cable cost: $720
- Connector cost reduction from elimination of indoor-outdoor connections: $7200

© Pearson Technologies Inc.

Fiber Cable Testing, Certification, And Troubleshooting — Chapter 4: Cables

Recommended Use

- When optoelectronics are more than 50' from entrance point
- When increased cable cost is less than savings from elimination of connections
  - Eliminated costs: connectors, enclosures, and labor

Bottom Line

- Specify UV resistance for outdoor cables
- Specify UV resistance and NEC compliance for indoor-outdoor cables
  - UV plus OFNR
  - UV plus OFNP

?? Remove 'resistances'??

High Voltage Resistance

- Definition: resistance to break down of plastics that leads to conductance
- Important when cable is installed in environment containing high voltage; e.g., around or near high voltage lines
- Example: 'Skywrap'
  - (AFL Telecommunications, http://www.afltele.com/products/fiber_optic_cable/skywrap/

Lightning Resistance

Lightning Resistance

- Definition: resistance to being damaged when struck by lightning
- Optical power ground wire (OPGW) is the answer
- OPGW cable can withstand 50,000-70,000 A with short term temperature increase to 300° C.
- High cost cable
- Used by power utilities

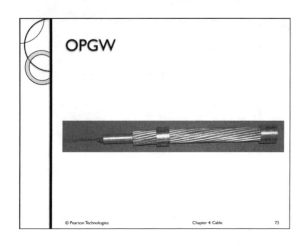

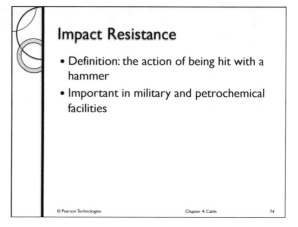

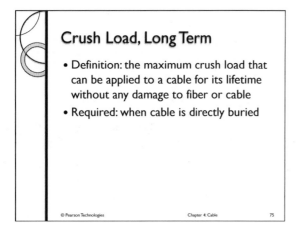

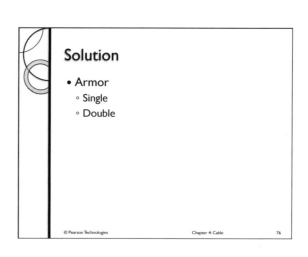

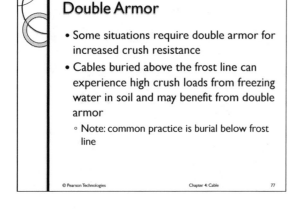

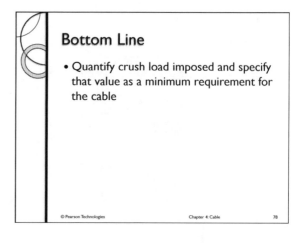

Fiber Cable Testing, Certification, And Troubleshooting — Chapter 4: Cables

Crush Load Resistance (Short Term)

- Definition: the maximum crush load that will be applied for a short period of time
- Required in
  - Broadcast and ENG applications
  - Field tactical military
  - Oil drilling rigs
  - Any environment in which cable can be walked on or driven over by heavy machinery

Ionizing Radiation Resistance

- Ionizing radiation: nuclear and x-rays
- Results in deteriorating optical and mechanical properties of cable
- Fiber attenuation increases
- Polymer flexibility decreases
- Important in nuclear power plants and in military applications (Cruise missile system)
- Specify dose and dose rate
- Radiation hard fibers and cables exist

Hydrogen Resistance

- Definition: increase in attenuation rate rate when fiber exposed to hydrogen gas
- Fiber attenuation increases
- Important in underwater cables which can corrode, releasing this gas
- Stainless steel used in such cables
- Alternatively, a hermetic coating is placed over cladding

Vibration Resistance

- In combination with other conditions, vibration in environment can result in increased attenuation rate
- One example: cables along freeway in S. California supporting *analog* CCTV (condition 1) monitoring system
- Traffic provided vibration (condition 2)

Vibration Situation 1

- Cables were loose tube, gel filled
  - Excess fiber length in cable (condition 3)
- S. CA has a desert environment (condition 4)
- Cold at nighttime, hot in daytime
- Cable not at same temperature along length
- Hot regions of cable expand into cool regions
- Excess fiber (aka coil) moves from hot to cool regions
- Attenuation in cool regions increases subtly

Vibration Situation 2

- Transmission distances were long: 15-20 km (condition 5)
- Long distances meant many locations in which attenuation rate increased
- Total increase in loss became significant
- Analog transmission requires a minimum signal to noise ratio
- S/N ratio dropped below requirement
- Tight tube cables replaced loose tube cables

SAP Cables And Vibration

- SAP cables may not exhibit this problem (Opinion)

Flame

- Certain environments require resistance to flame that is different from that in the NEC
- Such environments are: enclosed spaces and mobile platforms
- UL-94 defines test procedures and flame resistance from level V0 to V5

Gas Flow Resistance

- Definition: flow of gas through cable
- Important in nuclear reactors and underground nuclear test facilities
- Characteristic
  - Expensive cable
  - One of most difficult cables to manufacture

Filling Flow Resistance

- Definition: resistance to flow of filling materials
- Important in cable installations in which cable ends are at significantly different elevations
- Example: cable installed up the side of a mountain with a splice enclosure at bottom of mountain

Abrasion Resistance

- Required when environment will subject cable to significant abrasion
- Required in
  - Desert environments
  - Tunnels
  - Mines
  - Field tactical military applications
- Polyurethane is a common jacket material for high abrasion resistance

Flexing Resistance

- Definition: damage to or fiber cable materials when repeatedly flexed
- Standard performance cables allow flexing to 1000-2000 cycles
- Flexing resistant cables allow flexing to 10,000-20,000 cycles
- Both fibers and cable materials are improved
- Increased cost results

Fiber Cable Testing, Certification, And Troubleshooting — Chapter 4: Cables

Flexing Examples

- Undersea remotely operated vehicle
- Submarine detection sensors
- Cable on elevator car

Steam

- While operating temperature is an adequate specification for most situations, high steam levels can cause a problem
- Steam results in significant heat transfer to the medium on which it condenses, resulting in a jacket surface temperature increase that the environment does not see
- Jacket can melt off cable

Bottom Line

- Specify that cable must be resistant to exposure to steam

Chemical Resistance

- Cables must survive exposure to chemicals in the environment
- Examples
  - JP-4 at LAX
  - Lime in Canadian soil
  - Oil drilling rig in Gulf of Mexico

Bottom Line

- If environment contains chemicals that can attack the cable, cable must be resistant to degradation due to such exposure

Fiber Protection From Design

- Two groups of designs
  - Loose buffer tube
  - Tight buffer tube
- Four common designs

© Pearson Technologies Inc.

Fiber Cable Testing, Certification, And Troubleshooting Chapter 4: Cables

Buffer Tubes

- Definition: the buffer tube is the first layer of plastic placed around a fiber by the cable manufacturer

Two Types Of Buffer Tubes

- Loose tube, with a diameter of 2-3 mm or larger
- Tight tube, with a diameter of 0.9 mm (900 µ)

Loose Buffer Tube

- The inner diameter of the buffer tube is larger than the outer diameter of the fiber
- A loose buffer tube can contain one or more fibers
- 12 is the usual number of fibers
- 400 fibers is possible in a single, central, loose tube cable or a ribbon cable with a buffer tube diameter of 0.5"

Loose Buffer Tube Cross Section

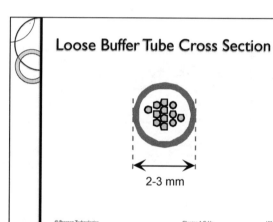

2-3 mm

Loose Buffer Tube Cable

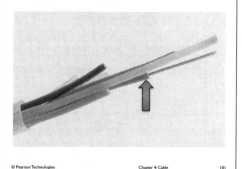

Tight Buffer Tube

- In a tight buffer tube, the inner diameter of the buffer tube is the same as the outer diameter of the fiber
- A tight buffer tube contains a single fiber

© Pearson Technologies Inc.

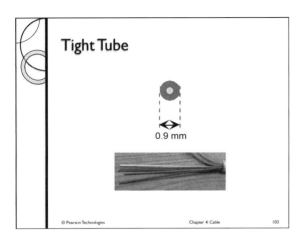
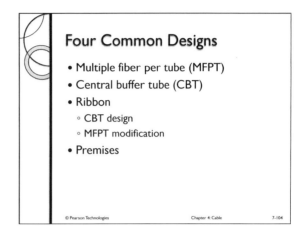
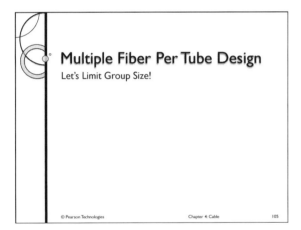
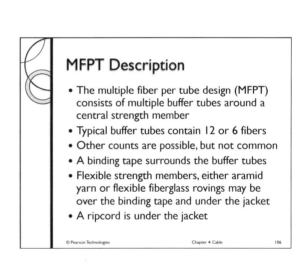
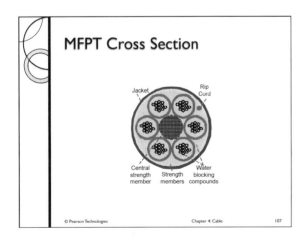
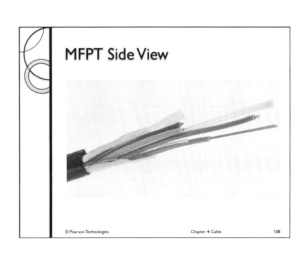

Fiber Cable Testing, Certification, And Troubleshooting — Chapter 4: Cables

Central Loose Buffer Tube Design (CBT)
12 is not the limit!

All Fibers In Central Buffer Tube

- There is no reason we need to limit the number of fibers to 6 or 12
- We can place all fibers in a single loose buffer tube
- The buffer tube will reside in the center of the cable, resulting in a design with the name 'central loose tube' or 'central buffer tube'
- In this design, fibers are in groups of 6 or 12, each group held together by a color coded thread or yarn
- Up to 216 fibers can reside in the buffer tube
- The buffer tube can be filled with water blocking gel or SAP.

CBT Structure

- Around the buffer tube are strength members, often flexible aramid yarns or fiberglass rovings
- The empty spaces outside the buffer tube can be filled with grease or with water blocking materials
- An outer jacket surrounds the strength members.
- As in the MFPT design, optional additional layers may include additional flexible strength members, armor, and a second jacket

Central Buffer Tube Cross Section
(7-7)

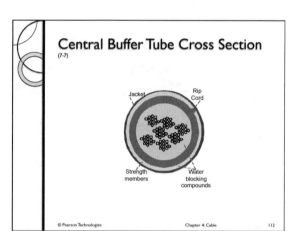

Ribbons

- A ribbon is a series of 4-24 fibers precisely aligned and glued to a thin tape substrate

12 Fiber Ribbon

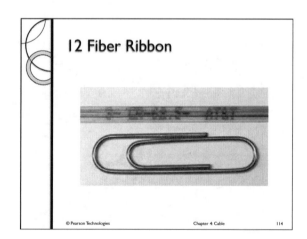

Ribbon Cable Description

- Ribbons can be stacked on one another to create a design with 416 fibers
- The ribbons are enclosed in the central loose buffer tube
- The buffer tube and the rest of the structure is the same as that of the central loose tube design
- An alternative ribbon cable design consists of ribbons stacked in loose buffer tubes, which are stranded around a central strengths member, as in the MFPT design

Ribbon Cable, CBT Version
(Courtesy Corning Cable Systems Inc.)

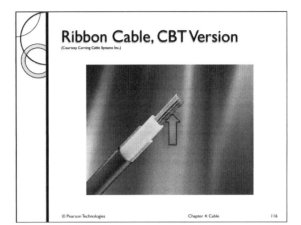

Ribbon Cable Cross Section

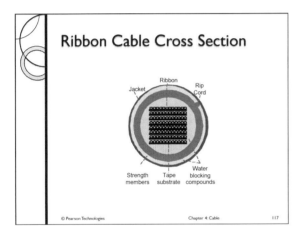

Ribbon Cable, Version 2

- A combination ribbons inside of multiple buffer tubes

Ribbon Cable, MFPT Version
(Courtesy Corning Cable Systems Inc.)

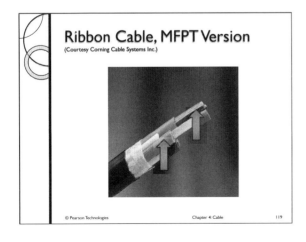

Design 4: Premises Cable

- The premises cable, aka 'distribution cable', is the most commonly used tight buffer tube design
- Commonly used in
  - Indoor networks
  - Field tactical, military applications
- Can be used in outside plant applications
- In COs, can connect loose tube OSP cable to indoor patch panel

Premises Cable Structure

- The premises cable structure consists of a centrally located strength member surrounded by a multiple, tightly buffered and stranded fibers
- Around the fibers is a layer of flexible strength members, usually aramid yarns, such as Kevlar®
- A jacket is extruded over the aramid yarns
- For moisture resistance, SAP tape or yarn is placed within the aramid yards

Premises Cable Cross Section

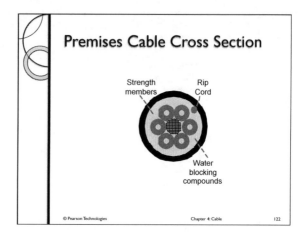

Limitation On Structure

- This structure used for up to 24 fibers
- For higher counts, this structure is repeated within an outer jacket

High Count Premises Cable

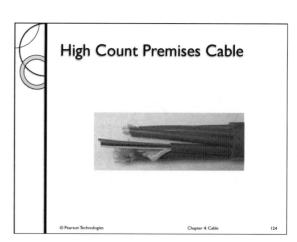

Cable Colors

- Yellow- indoor singlemode
- Aqua- indoor LO, 50µ
- Orange- generally, indoor 62.5µ
- Gray- generally, indoor 50µ

Fiber Optic Test Procedures

- All these characteristics require performances defined by a specific fiber optic test procedure
- Test procedures are known as FOTPs
- Most test procedures are defined with two designations
  - FOTP-xx
  - EIA/TIA-455-xx
- Some test procedures are defined in other test standards

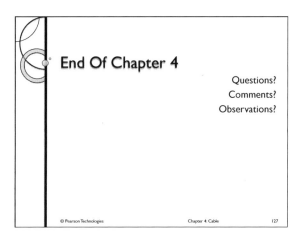

Fiber Cable Testing, Certification, And Troubleshooting   Chapter 5: Connectors

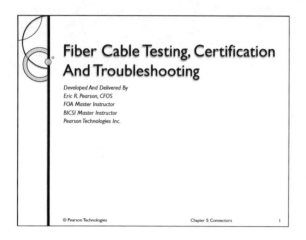

Fiber Cable Testing, Certification And Troubleshooting
Developed And Delivered By
Eric R. Pearson, CFOS
FOA Master Instructor
BICSI Master Instructor
Pearson Technologies Inc.

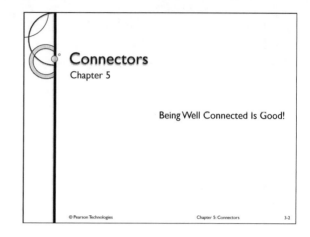

Connectors
Chapter 5

Being Well Connected Is Good!

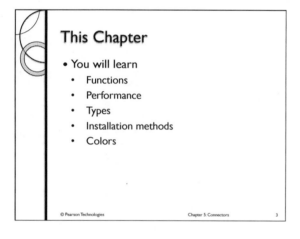

This Chapter
- You will learn
  - Functions
  - Performance
  - Types
  - Installation methods
  - Colors

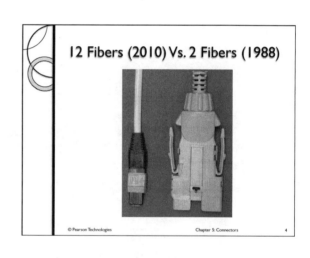

12 Fibers (2010) Vs. 2 Fibers (1988)

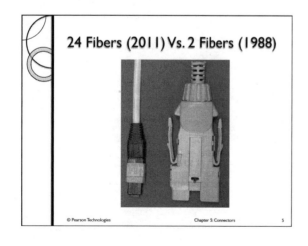

24 Fibers (2011) Vs. 2 Fibers (1988)

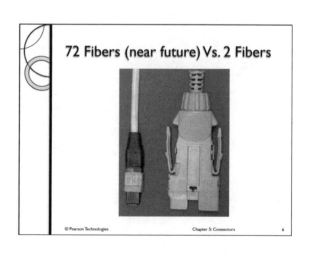

72 Fibers (near future) Vs. 2 Fibers

Fiber Cable Testing, Certification, And Troubleshooting   Chapter5: Connectors

Benefits
- You will be able to
  - Understand connector performance requirements
  - Recognize connector type
  - Evaluate connector installation methods

Four Functions
- Low power loss
- High fiber retention strength
- End protection
- Disconnection

Two Connector Performances
- Optical
- Environmental

Three Optical Performances
1. Maximum insertion loss, dB/pair
2. Typical insertion loss, dB/pair
3. Reflectance, in – dB

Maximum Insertion Loss
- Units "dB/pair"
- A measurement of power loss from one fiber to another
  - Must be "dB /pair"
- Is the maximum loss the installer will experience with correct installation
- Designers specify this value
- Installers use this value in certifying networks

How High The Loss?
- With incorrect installation, there is no limit to the power loss!

Fiber Cable Testing, Certification, And Troubleshooting   Chapter5: Connectors

Consequence Of dB/Pair

- No power loss from a transmitter to a fiber or from a fiber to a receiver

A Magic Value

- Maximum insertion loss = 0.75 dB/pair

Two Types Of Causes Of Loss

- Intrinsic
- Extrinsic

Intrinsic Loss

- When the installer installs connectors correctly, the loss is due to intrinsic causes,
  - Core and cladding diameter variations
  - Core offset
  - Cladding non-circularity
  - NA mismatch
  - Offset of the fiber in the ferrule

Extrinsic Loss

- When the installer makes errors, he introduces extrinsic causes of loss
  - Damaged core
  - Bad cleaves
  - Dirt or contamination on the core
  - Air gaps due to excessive polishing

Optical Performance Concerns

1. Maximum insertion loss, dB/pair
2. Typical insertion loss, dB/pair
3. Reflectance, in – dB

© Pearson Technologies Inc.

Typical Insertion Loss

- Do not expect to see the maximum value
- The typical insertion loss, in dB/pair, is the value the connector will have when properly installed
- Frequently, the connector will have a value lower than this value
- The designer and the installer use this value in certifying networks
- You want to know this value, but it is not a specification

Three Typical Loss Values

- Polished connector with 2.5 mm ferrule
  - = 0.30 dB/pair
- Polished connector with 1.25 mm ferrule
  - = 0.20 dB/pair
- Pre-polished connector with 2.5 mm ferrule
  - = 0.40 dB/pair

Optical Performance Concerns

1. Maximum insertion loss, dB/pair
2. Typical insertion loss, dB/pair
3. Reflectance, in − dB

Reflectance 1

- A measurement of the relative optical power reflected backwards from a connector
- Reflected power can travel back to the light source and be reflected from the source back into the fiber
- If this reflected power reaches the receiver with a power level above its sensitivity, the receiver will convert this optical power to a digital 'one'

Reflectance 2

- If the time interval in which the reflected power arrived was a digital 'zero' at the transmitter, the output signal will differ from the input signal
- Reflectance influences the accuracy with which a fiber optic link will transmit
- Minimizing reflectance results in maximizing signal accuracy
- The objective of a reflectance requirement is to limit the reflected power at the receiver to a value less than the sensitivity of that receiver

Reflectance = Fresnel Reflection

- Reflectance occurs at a glass air interface
- This interface, or any interface at which there is a change in speed of light, or index of refraction, can produce a reflection, called a "Fresnel reflection"
- This reflection occurs in fiber optic connectors because the end faces of mated connectors have surface roughness, which creates microscopic air gaps

Air Gaps Create Reflectance

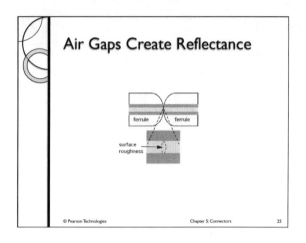

Reflectance Definition and Range

- Definition
  - 10 log (reflected power/incident power)
- Reflectance is stated in units of negative dB
  - Range from -20 dB to -65 dB

PC, UPC And APC

- Reflectance is qualitatively described by
  - PC (physical contact)
  - UPC (ultra physical contact)
  - APC (angled physical contact)
- PC < –40 dB
- UPC < –50 dB
- APC < –55 dB

Typical Reflectance Values

- The Fresnel reflection of fiber to air
  - -14 dB to -18 dB (from a non-contact connector)
- From hand polishing singlemode connector
  - < -50 dB
- From factory polished singlemode connector
  - < -55 dB
- APC
  - < -65 dB

Recommendation

- Whenever reflectance is of concern, fusion splice factory-polished singlemode pigtails
  - Instead of field installation of connectors

Two Low Reflectance End Faces

- Radius ferrules
- APC ferrules

Radius End Face Ferrules

- Eliminate the need to achieve perfect perpendicularity
- With a end face radius, reflectance can be reduced through multiple polishing steps with successively finer polishing films
- The principle
  - The smoother the surface, the smaller the air gaps and the fewer the air gaps
  - The smaller the air gaps and the fewer the air gaps, the lower will be the reflectance

End Faces With Radius

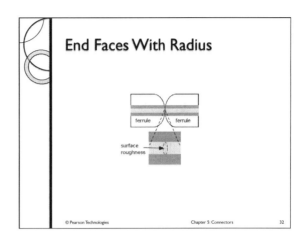

APC Reflectance

- Angled Physical Contact (APC) connectors provide reflectance values better than (less than) -60 dB
- My experience: APC < -64 dB
- APC connectors are green

APC End Face Ferrules

- Tip of the ferrule is at 8 degrees to the perpendicular of the fiber axis
- Angle reflects light backwards at an 8° angle to the axis, which is outside of the critical angle of singlemode fibers
- All reflected light escapes from the core
- No light returns to input end

APC Versions

- LC
- LX.5
- SC
- FC

SC/APC

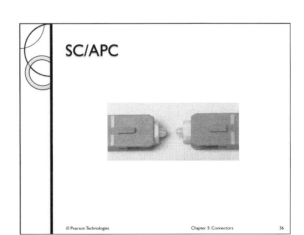

Fiber Cable Testing, Certification, And Troubleshooting — Chapter5: Connectors

---

TIA/EIA-568-C Reflectance

- TIA/EIA-568-C allows
  - A multimode connector to increase in reflectance from its initial value to -20 dB after 500 cycles
  - A singlemode connector to increase in reflectance from its initial value to -26 dB after 500 cycles
  - Per FOTP-21
  - Source: TIA/EIA-568-C.3, Annex A.4.9

Commonly Confused Terms

- Return loss and back reflection
- Return loss is the power reflected backwards by a system or a link
  - Return loss includes reflectance and the backscatter from atoms in the core of the fiber
- Back reflection applies to all components in a link
- Back reflection is defined as the reciprocal of reflectance and is, therefore, a positive number

---

Connector Types

- Commonly used types
  - Dominant at this time
- Small form factor (SFF)
  - SFF types are expected to become dominant in the future
- Legacy types
  - Were used in the past but are no longer used for initial installations

---

Two Common Types

- ST™-compatible
- SC

  - ST is a trademark of Lucent Technologies

Common Characteristics

- Keyed
- Contact
- Moderate loss

ST-Compatible

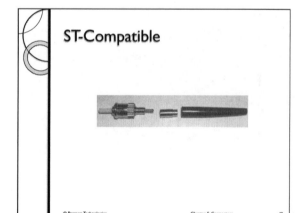

© Pearson Technologies Inc.

ST-Compatible Origin

- Introduced in 1986
- Installation is significantly easier and faster than that of predecessor, or legacy, types

ST-Compatible Characteristics

- Simplex only
- Not pull proof
- Not wiggle proof
- Require large spacing
  - 12 connectors in 1U enclosure

SC Simplex Connector

SC Origin

- Available in 1988
- Designed by Nippon Telephone and Telegraph (NTT) in Japan
- Became commonly used when it became the connector type required for compliance with the Building Wiring Standard, TIA/EIA-568

SC Characteristics 1

- Simplex
- Can be made duplex
- Pull proof
- Wiggle proof
- Require reduced spacing
  - 24-48 in 1U enclosure

SC Duplex

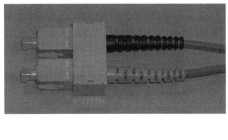

SC Characteristics 2

- Less expensive to use than the ST-compatible
- More expensive to buy than the ST-compatible
- More reliable than the ST-compatible

SC Cost Advantage

- Reduced spacing reduces total cost of SC connectors and enclosures
  - By reducing the cost, number, or size, of enclosures required

SC Reliability

- Immune to damage from 'pull and snap'
- An SC connector cannot be pulled and allowed to snap back into an adapter
  - An ST-compatible connector can be pulled and snapped, usually, resulting in fiber damage
- Damage: ST-compatible reference leads frequency is 3-6 times that of SC reference leads
- Life cycle cost of the SC connector is lower than that of the ST-compatible connector

Small Form Factor Connectors
Bigger Is Not Better!

Time Shrinks Connectors

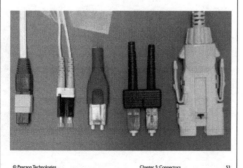

24 Fibers (2011) -2 Fibers (1988)

ST And SC Disadvantage

- ST-™ compatible and SC are large connectors
- The large size forces optoelectronics manufacturers to space transmit and receive electronics far apart
- This large spacing results in half as many fiber ports in a fiber hub or switch as in a UTP hub or switch
- Size increases optoelectronics cost

Small Form Factor Connectors

- Optoelectronics manufacturers requested development of a 'small form factor' (SFF) connector

SFF Advantage

- Reduced the per port cost of hubs and switches
- Goal was for cost parity of fiber switches with UTP switches
- Connector manufacturers answered this request with a series of small form factor (SFF) connectors

Size Reduction Obvious

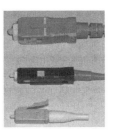

Two Common SFF Connectors

- LC
- MTP/MTO
- There are five others, but they are not common in telco or cell phone tower networks

SFF Common Characteristics

- Keyed
- Contact
- Push on- pull off

LC Simplex

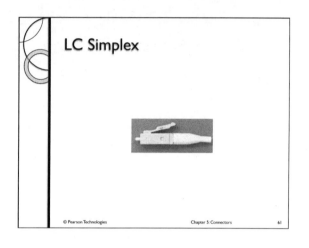

LC Background

- Available 1997-1998
- Developed by Lucent Technologies
- Available from other manufacturers
- Developed as a telephone connector designed to increase the density of installed connectors
- DWDM is one of the technologies creating the need for increased density
- A simplex SFF connector that can be converted to a duplex form with a clip

LC Duplex

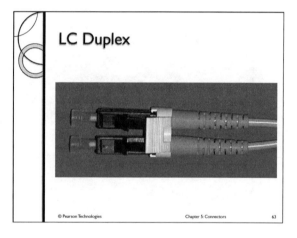

LC Characteristics

- Low loss: typical 0.15-0.20 dB/pair
- Pull proof
- Wiggle proof design
- New: 1.25 mm ferrule
- Many optoelectronics have the LC interface
- Preferred interface for 10Gbps optoelectronics (opinion)

Recommendation

- Choose LC
- Reasons
  - Low loss
  - 16 years of use has revealed no hidden problems
  - Available from multiple manufacturers
  - Available in multiple installation methods

LX.5

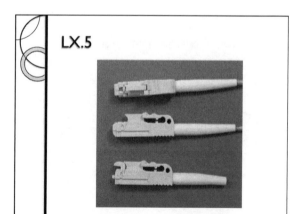

LX.5 Unique Features

- Built in dust covers in
  - Connector
  - Barrel
- Act as dust covers and as eye safety shutters

LX.5 Built In Ferrule Cover

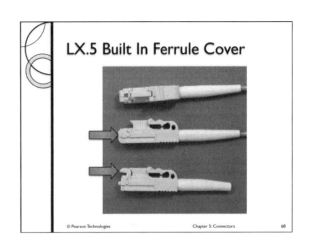

LX.5 Built In Barrel Shutter

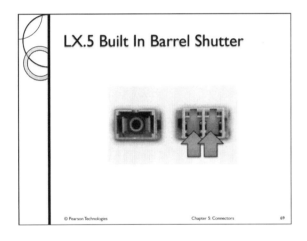

Note

- Doubling of connector density: the left barrel is a simplex SC barrel

Observation

- LX.5 was developed for telephone industry
- This author has never seen or heard of it being used in a data communication system

MU

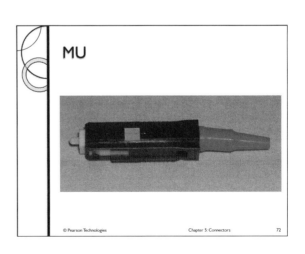

MU Background

- The appearance of an SC but with all dimensions reduced by 50 %
- Designed and developed in Japan
- Used in the US by Alcatel in FTTH systems
- Made in the US under license by Tyco/AMP

E2000

- European connector
- Built in dust cover
- Similar to LX.5
- 1.25 mm ferrule

E2000 Connector

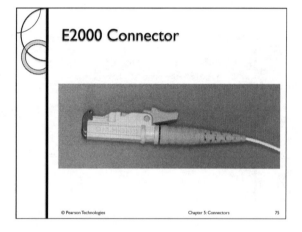

MTP/MPO

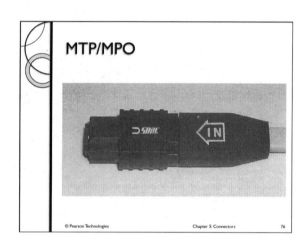

MTP/MPO Background

- MPO: Multiple-Fiber Push-On/Pull-Off
- Used in pre-terminated systems
  - Indoors in data centers
  - Outdoors in FTTH networks
- Will be used in 40 and 100 Gbps Ethernet networks
- Ribbon connector
  - 12 and 24 fiber versions available
  - Can be used with 12 fiber/tube, loose tube cables
    - MTP® is a registered trademark of US Conec Ltd.

MTP/MPO: 12 & 24 Fibers
(Photograph Courtesy Molex)

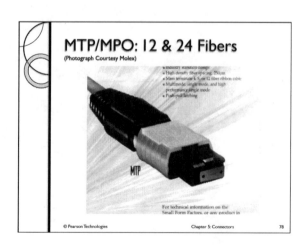

Fiber Cable Testing, Certification, And Troubleshooting — Chapter 5: Connectors

40 and 100 Gbps Ethernet

- Will de-multiplex the data stream into 10 Gbps streams
- 40 Gbps will require 4 fibers for transmit and 4 fibers for receive
- 100 Gbps will require 10 fibers for transmit and 10 fibers for receive
- Low skew becomes critical

40 And 100 GB Ethernet Multimode Transmission

- Will be accomplished with MTP/MTO connectors
- Ideal solution

MTP/MPO Characteristics

- Typical loss: ≤0.5 dB/pair, multimode
- Typical loss: ≤0.75 dB/pair, singlemode
- Available in radius and APC versions
- Connector size approximately same as that of SC

MTP– SC Size Comparison

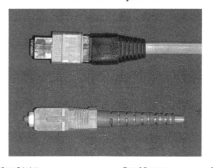

Advantages

- Reduced installation cost
  - Data centers
  - Riser networks
- Reduced network cost
  - Reduced patch panel cost and space

Legacy Connectors

- We shall skip these, as they are not used in new networks

Fiber Cable Testing, Certification, And Troubleshooting   Chapter5: Connectors

Installation Methods Data Base

- Basis for this next section is my experience in training in and field connector installation
- This experience includes
  - More than 7416 installation trainees
  - More than 46,140 connectors installed in training or the field
- Methods include: epoxy, quick cure adhesive, Hot Melt™, no adhesive and polish, no adhesive and no polish, pigtail

Two Concerns

- Environment compatibility
- Total installed cost

Environment Compatibility

- Heat cure in cold environment?
- Polish in windy, dusty, or wet environments?

Total Installed Cost

- Not connector cost, but total installed cost
- Low cost connectors can have high total installed cost
- High cost connectors can have high total installed cost
- Total installed cost includes
  - Connector cost
  - Labor cost
  - Process yield

Six Methods

1. Epoxy
2. Hot Melt™
3. Quick Cure Adhesive
4. 'Cleave-and-Crimp' Installation
5. Fusion spliced pigtails
6. Fuse on connectors

Review Two Methods

- Hot Melt Adhesive
  - Demonstration
- 'Cleave-and-Crimp'/No Polish Installation
  - Hands-On Activity
- Fusion spliced pigtails
- Fuse on connectors

© Pearson Technologies Inc.

Fiber Cable Testing, Certification, And Troubleshooting Chapter5: Connectors

Recommended Methods

- Fusion spliced pigtails
- Fuse on connectors
- For fastest restoration
  - Cleave and Crimp

Fusion Spliced Pigtails

- The connector installation method with the lowest total installed cost does not require connector installation
  - Fusion splice factory installed pigtails
- A pigtail is a short length of jacketed cable or 900 µ fiber with a connector on one end
- Cut a patch cord in half to create two pigtails

The Benefit

1. 100 % yield
2. High reliability
3. Low reflectance

Details

- The reduction in cost will pay for a low cost fusion splicer
  - 725-3000 connectors (Pearson Technologies Inc. analysis)
- Cost factors
  - Pigtail at $3-5
  - Splice tray at $1/connector
  - Splice cover at $0.40/splice
  - Splice labor at 12-20 splices per hour and 98% yield
- See http://www.ptnowire.com/tpp-V3-I2.htm

Splice/Fuse On Connectors

- Use a fusion splicer with adapter kit
- Connector cost $12-$20
- Ideal for retro-fitting when space for splice trays not available in enclosure or rack
- High yields, but not 100 %

Connector Colors

- Connectors, boots, and barrels are color coded
- Green: APC (singlemode)
- Blue: singlemode
- Beige: 62.5 µ multimode
- Aqua: 50 µ laser optimized
- Black: 50 µ
  - TIA/EIA-568-C.3, A 4.2.3

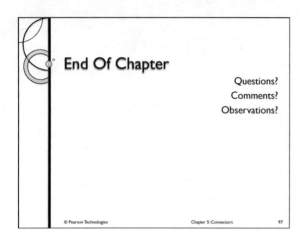

Fiber Cable Testing, Certification, And Troubleshooting   Chapter 6- Connector inspection

Fiber Cable Testing, Certification And Troubleshooting

Developed And Delivered By
Eric R. Pearson, CFOS
FOA Master Instructor
BICSI Master Instructor
Pearson Technologies Inc.

Connector Inspection
Chapter 20

What You See Is What You Get!

This Chapter

- Learn how to inspect connectors
- Learn how to rate or interpret microscopic appearances
- Learn how to identify corrective actions

Applicability

- This chapter applies
  - To all connectors that require field polishing
  - To connectors that have been in use
- We assume factory polished connectors are correct

Reason For Inspection

- A 'good' appearance correlates with low loss, most of the time
- Cleave and crimp connectors may not exhibit this correlation, because loss is determined by quality of cleaved fiber ends inside the connectors

Equipment Required

- A 400-magnification connector inspection microscope with an IR filter
  - Some professionals recommend 100x and 200x microscopes
  - Recommended: www.Westoverscientific.com
- Lens grade, lint free, tissues (Kim wipes or equivalent)
- Electro-Wash® Px (recommended)
  - Or 98% isopropyl alcohol

© Pearson Technologies Inc.

Fiber Cable Testing, Certification, And Troubleshooting   Chapter 6- Connector inspection

Procedure
- Remove cap
- Clean connector
- Install connector
- Focus
- View and rate connector
- Repeat viewing and rating with back light

Back Light
- Is a white light on opposite end of fiber
- Back light can reveal features
  - Example: fiber broken below surface but in ferrule
- Back light can conceal features
- Whenever possible, inspect both ways

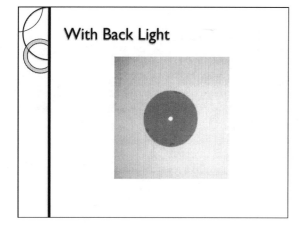

With Back Light

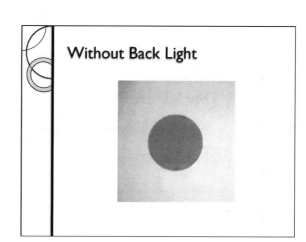

Without Back Light

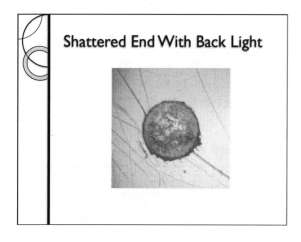

Shattered End With Back Light

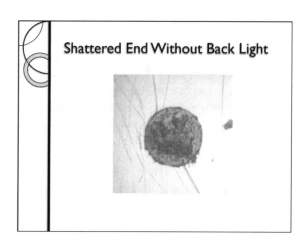

Shattered End Without Back Light

© Pearson Technologies Inc.

Fiber Cable Testing, Certification, And Troubleshooting   Chapter 6- Connector inspection

Key Facts
- Light travels in core
- Connectors are contact

Evaluation Criteria
- Core
  - Round
  - Clear
  - Featureless
  - Flush
- Cladding and ferrule surface
  - Clean

Good Core With Back Light
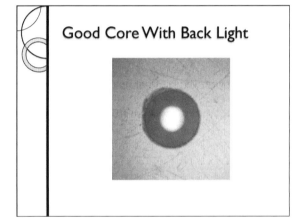

Cleaning Residue
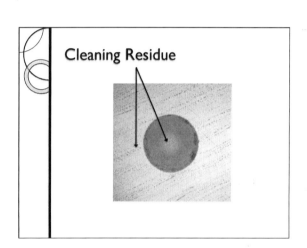

Cleaning Residue
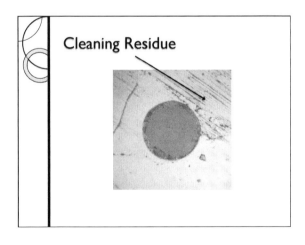

Scratch/Feature On Core
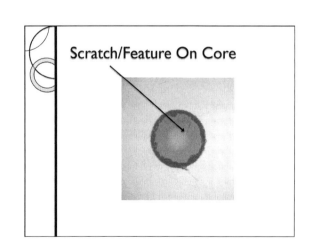

Fiber Cable Testing, Certification, And Troubleshooting   Chapter 6- Connector inspection

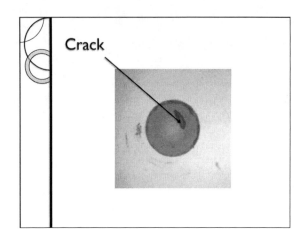
Crack

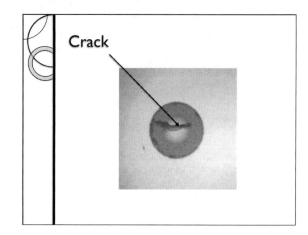
Crack

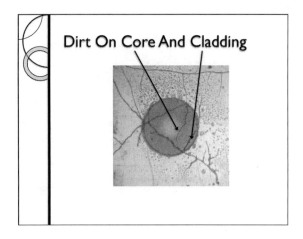
Dirt On Core And Cladding

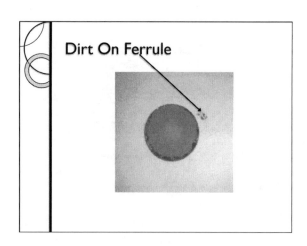
Dirt On Ferrule

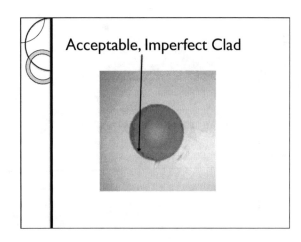
Acceptable, Imperfect Clad

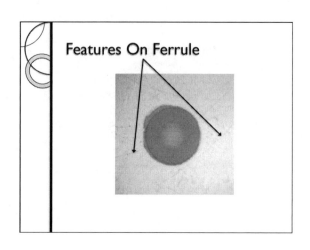
Features On Ferrule

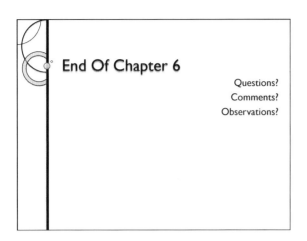

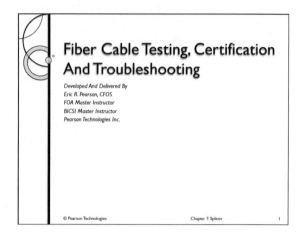
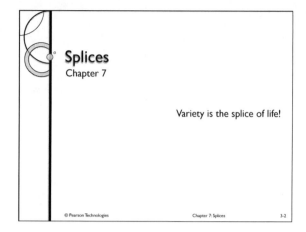
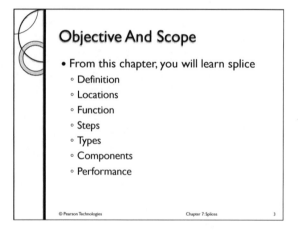
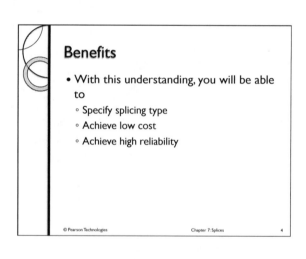
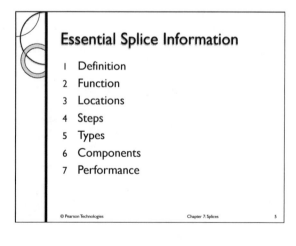
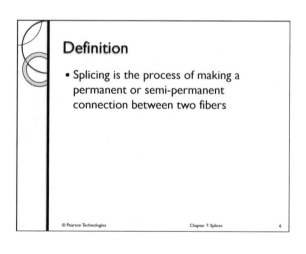

Fiber Cable Testing, Certification, And Troubleshooting   Chapter 7: Splices

Essential Splice Information
1. Definition
2. Locations
3. Functions
4. Steps
5. Types
6. Components
7. Performance

Two Splice Functions
- Low loss
- High strength

Essential Splice Information
1. Definition
2. Locations
3. Function
4. Steps
5. Types
6. Components
7. Performance

Two Locations And Reasons
- Mid Span
  - The cable cannot be obtained in a single length
  - The cable cannot be installed in a single length
  - The cable has experienced back hoe fade, post hole driller fade, rodent fade, shark fade, drunk driver fade, or any other form of failure
- Pigtail
  - Pigtails are short length of cables or tight buffer tubes with a connector installed on one end
  - Installers perform pigtail splicing in order to reduce installation cost or installation time
  - Installers install the majority of singlemode connectors by pigtail splicing

Splice Locations
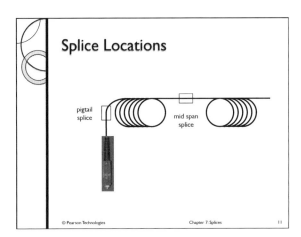

Essential Splice Information
1. Definition
2. Locations
3. Function
4. Steps
5. Types
6. Components
7. Performance

Splicing Steps

- Prepare cable ends
- Prepare two fiber ends
- Clean fibers
- Cleave fiber ends
  - Cleaving is creation of nearly perfectly perpendicular and smooth ends
- Make splice
- Route fibers in tray
- Place splice in tray
- Place tray in enclosure
- Close and seal enclosure

Essential Splice Information

1. Definition
2. Locations
3. Function
4. Steps
5. Types
6. Components
7. Performance

Two Types

- Fusion
- Mechanical

Tools Required

- Almost the same tool kit for both methods
  - Base kit cost $2000-$2500
  - Most of cost is for high precision cleaver ($1300-$1600)
- Fusion splicing requires fusion splicer
- Mechanical splicing may require an inexpensive tool

Fusion Splicing

- Is the process of fusing, or welding together, of two fibers
  - Aka glass welding

Fusion Splicing Used In

- Most initial installations
- In restoration
- Requires fusion splicer

Fiber Cable Testing, Certification, And Troubleshooting   Chapter 7: Splices

Two Splicer Functions

- The precise alignment of the fibers to each other prior to splicing
- Precise control of the splicing operation

Precise Alignment

- Precise alignment means sub micron precision
- Alignment results in low power loss

Two Alignment Methods

- Passive
- Active
- Costs
  - Passive alignment: $5,000-$7,000
  - Active alignment: $8,000-$30,000
    - At ≥ $8,000, times are
      - Make a splice in 9 seconds
      - Shrink the splice cover in 35 seconds

Passive Alignment

- Based on a precision 'V' groove
- Assumptions
  - Fiber diameters and the
  - Core-cladding concentricity are precise enough to achieve low power loss
- These assumptions are valid for both the fiber made in North America and much (but not all) of the fiber made overseas

Active Alignment

- Active means fibers are moved to provide low loss
- Two methods
  - Profile alignment (aka PAL and PAS)
    - Dominant method
    - From Fujikura, Sumitomo, Fitel
  - Local injection and detection (LID)
    - From Corning Cable Systems

Two Active Alignment Methods

- Profile alignment (aka PAL and PAS)
  - Dominant method
  - From Fujikura, Sumitomo, Fitel
- Local injection and detection (LID)
  - From Corning Cable Systems

Fiber Cable Testing, Certification, And Troubleshooting — Chapter 7: Splices

Both Methods Provide
- Low loss
- High strength

PAS
- Collimated light is passed through both fibers
- Collimated light reveals core-cladding boundary
- Microscope lens behind fibers collects image
- Image is digitized
- Core cladding boundary is identified
- Fibers are move in X and Y axes to minimize offset of cores

PAS Alignment Mechanism

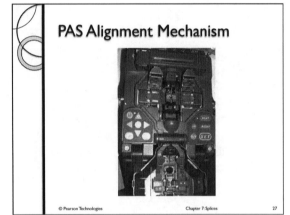

PAS Final Screen

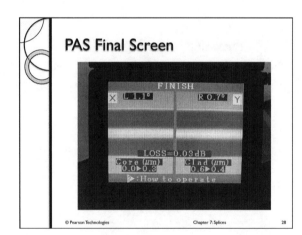

LID
- Light is launched into one fiber and tapped from the second fiber
- Fibers are moved to provide maximum power transfer between two fibers

Precision Control
- The splicer aligns the fibers
- Splicer controls the
  - Arc current
  - Arc time
  - Overrun
- These parameters determine the power loss and strength of the splice

Fiber Cable Testing, Certification, And Troubleshooting — Chapter 7: Splices

---

Fusion Splicing Advantages
- Low power loss
  - Often, fusion splices result in 0 dB loss
- Low to no reflectance
  - Theory says 'low' reflectance (due to different IRs)
  - In 36 years, I have never seen reflectance on a properly made fusion splice
- Low cost per splice
  - Only unique consumable is fusion splice cover @ < $1/splice

---

Two Fusion Splicing Disadvantages
- Fusion splicer can be expensive
- Fusion splicing may be cost prohibitive, even with a rented splicer if the number of splices is low
  - $2000 rental charge for 100 splices= $20/splice
  - A mechanical splice will be less expensive

One Final Disadvantage: No Manhole Fusion Splicing

-manholes collect methane!

Mechanical Splicing
- Is the process of placing two cleaved ends in a mechanical splice
- Splice provides cladding alignment and retention, or, strength
- The mechanical splice provides the function of a cover

Mechanical Splice Types

Alignment Differences
- Some splices have a precision capillary tube for alignment
- Other splices use a precision etched silicon substrate
- The 3M FibrLok2® has a precision 'V' groove

Mechanical Splice Common Feature

- All mechanical splices have index matching gel at their centers
- Gel fills any air gap that may result from the end faces of being less than perfectly perpendicular (which they are)
- Such filling reduces loss
- In some singlemode fibers, the gel eliminates all reflectance
  - When all three indices are the same

Mechanical Splicing Used

- Organizations that have a large number of splice teams
  - Reduced capital equipment cost
- Small number of splices
  - Reduced cost per splice
- Old multimode fibers
  - No disruption of core profile
- For fast restoration
  - Must have a fast restoration kit

Fast Restoration Kit

- Is a length of cable, of the same type as is in the network
- Both ends are prepared and installed in two enclosures
- All fibers are prepared and installed in mechanical splices
- The splices are in trays
- Trays are in the enclosures
- Enclosures and cable are in a large suitcase

Fast Restoration Kit Use

- When service loop cannot be pulled into area of repair to provide cable length for repair
  - Loop inserted between broken ends
  - Two splices per fiber required
- When fast restoration is required

Advantage

- Reduces restoration time by 50 %
- Half of preparation has been performed ahead of time

Recommendation

- Use FibrLok (p/n 2529) from 3M™
- Consistent low loss
  - Typical loss, singlemode or multimode, of 0.05 dB
  - Even with mismatched fibers
  - 2529 is singlemode and multimode!
- Opinion

Fiber Cable Testing, Certification, And Troubleshooting Chapter 7: Splices

Advantages

- Low capital equipment cost
- Many are re-enterable or reusable
- Some can be tuned to low loss with VFL

Mechanical Splice Disadvantage

- Major disadvantage
  - High cost in large installations
  - At a price of $8-$15, a 1000 splice installation will cost $8000-$15,000
- At $9,000, a fusion splicer becomes attractive alternative

Potential Disadvantage

- Increased installation time required to 'tune' splice to low loss
- This disadvantage occurs in some, but not all, mechanical splices

Gripping Details

- All mechanical splices grip by compression on the fiber
  - Some grip with drill chuck type mechanism
  - Some grip be closing spring loaded mechanism

Essential Splice Information

1. Definition
2. Locations
3. Function
4. Steps
5. Types
6. Components
7. Performance

Three Primary Splice Components

- Splice and cover
- Splice tray
- Splice enclosure

Fiber Cable Testing, Certification, And Troubleshooting   Chapter 7: Splices

Splice Cover

- The installer places a splice cover over the fusion splice
- The cover isolates the splice from the environment, and supports and protects the splice
- Splice covers can be heat shrinkable or adhesive

Fusion Splice Covers
(Courtesy of Preformed Line Products)

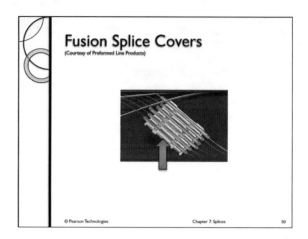

Adhesive Splice Covers

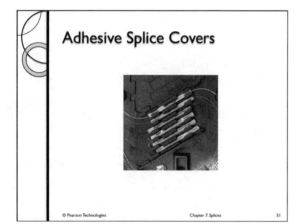

Splice Tray

- The tray protects the splice and houses excess fiber

Full Length Splice Tray

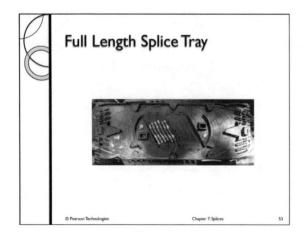

Half Length Splice Tray
(Courtesy of Preformed Line Products)

Splice Enclosure

- The enclosure houses the trays and excess buffer tube

Outdoor Splice Enclosure
(Courtesy of Preformed Line Products)

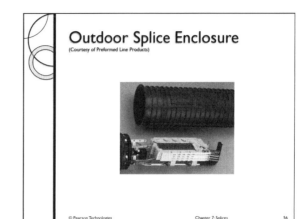

Indoor Splice Enclosure

Enclosure

- The enclosure houses and protects the splice trays and excess buffer tube length
- An inside enclosure can include an internal patch panel, complete with barrels
- Excess buffer tube allows the splice tray to be outside of the splice enclosure while the installer makes the splice
- While not a rule of thumb, outdoor splice enclosures require four to five feet of buffer tube per cable per end

Enclosure Functions

- Grip the cable strength members
  - So the cable cannot be easily pulled from the enclosure
- Provide space for splice trays
  - Not all indoor enclosures do so
- Isolate of the interior from the environmental conditions, such as moisture and dust
  - Unlike outdoor enclosures, indoor enclosures rarely provide such isolation

Secondary Components

- Moisture seals
- Gaskets
- Strength member gripping mechanisms
- Pressure valves
- Internal, or integral, patch panels
- Grounding strips
- Locking mechanisms
- Plugs

Indoor Enclosure Gripping Mechanism

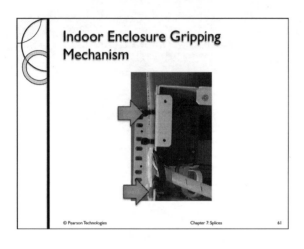

Outdoor Enclosure Gripping Mechanism

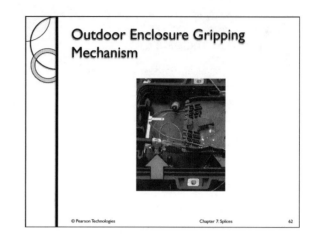

Moisture Isolation

- Some outdoor, aerial enclosures do not provide complete moisture isolation
- Instead, they provide weep holes so that condensed moisture can drain from the enclosure

Moisture Resistance

- Some outdoor enclosures provide mechanism for internal air pressure to prevent moisture entrance
  - Required when installed below ground level
- Such enclosures have a complex set of gaskets to prevent moisture ingress or internal pressure release

Valve For Internal Pressure

Integral Patch Panels

- Unlike most outdoor enclosures, indoor enclosures include an internal patch panel for direct connection of connectors to the fibers in the enclosure

Fiber Cable Testing, Certification, And Troubleshooting — Chapter 7: Splices

Essential Splice Information

1. Definition
2. Locations
3. Function
4. Steps
5. Types
6. Components
7. Performance

Performance

- Splice loss is less than 0.15 dB, though some organizations allow higher or require lower values
- Both fusion and mechanical splices achieve this value, although fusion splices tend to have lower loss than mechanical splices

Splice Loss Options

- The Fiber Optic Association advanced certification process requires all splices to be a maximum of 0.15 dB
- The Building Wiring Standard, TIA/EIA-568 C, allows splices to be up to 0.3 dB
- Some organizations allow splices to be as high as 0.5 dB, probably for cost reasons

Splice Specification

- Type
- Cladding diameter
- Mode type
- Maximum loss, in dB

End Of Chapter 7

Questions?
Comments?
Observations?

Fiber Cable Testing, Certification, And Troubleshooting — Chapter 8: Passive Devices

Fiber Cable Testing, Certification And Troubleshooting

Developed And Delivered By
Eric R. Pearson, CFOS
FOA Master Instructor
BICSI Master Instructor
Pearson Technologies Inc.

Passive Devices
Chapter 8

Passive Devices

- Modify optical signal in optical regime, without converting signal to an electrical signal
- Usually do not require power
  - Exceptions: optical amplifiers, optical switches
- Benefits
  - Reduce cost
  - Increase flexibility

Functions

- Couplers- combine different wavelengths onto same fiber; multiplexing
- Demultiplexing- separating wavelengths
- Splitters- transmit same signal to multiple receivers
- Switches- move optical signal from one path to another

Uses

- Telephone networks
  - CWDM
  - DWDM
- FTTH networks
- POLAN networks
- Not in data networks

Performance Concerns

- Power loss
- Reflectance

Fiber Cable Testing, Certification, And Troubleshooting   Chapter 8: Passive Devices

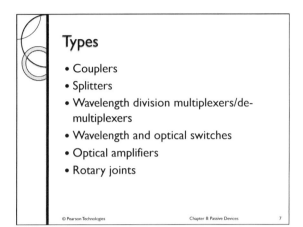

Types
- Couplers
- Splitters
- Wavelength division multiplexers/de-multiplexers
- Wavelength and optical switches
- Optical amplifiers
- Rotary joints

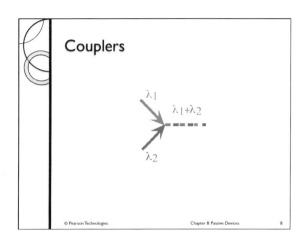

Couplers

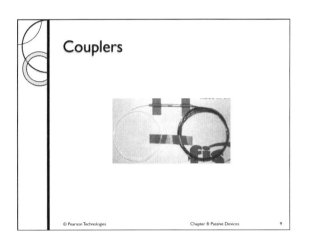

Couplers

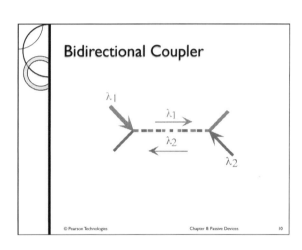

Bidirectional Coupler

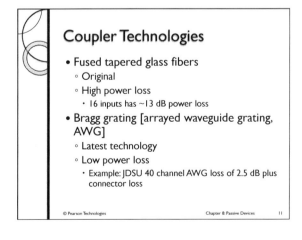

Coupler Technologies
- Fused tapered glass fibers
  - Original
  - High power loss
    - 16 inputs has ~13 dB power loss
- Bragg grating [arrayed waveguide grating, AWG]
  - Latest technology
  - Low power loss
    - Example: JDSU 40 channel AWG loss of 2.5 dB plus connector loss

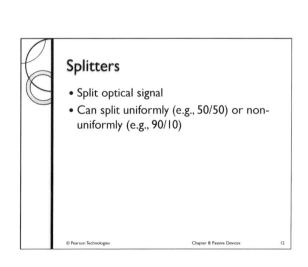

Splitters
- Split optical signal
- Can split uniformly (e.g., 50/50) or non-uniformly (e.g., 90/10)

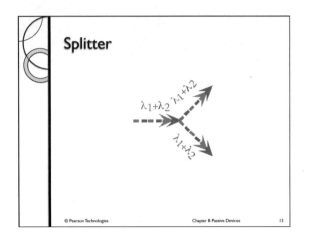

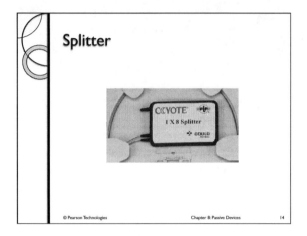

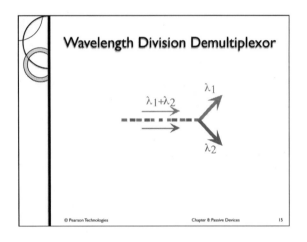

De-multiplexer Technologies

- Fused fiber splitter
  - Original technology
  - High power loss
- AWG [arrayed waveguide grating]
  - Latest technology
  - Low power loss
  - Coupler can be configured as de-multiplexer

Switches

- Use micro mirrors to reroute optical signals

ROADM

- Reconfigurable optical add/drop multiplexer
- Enables dropping and dropping wavelengths mid-span
- Can be remotely configured

ROADM [Courtesy NTT]

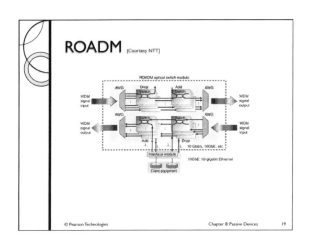

Optical Amplifiers

- Amplify optical signal
- Operate at 1550 and 1310 nm
- Amplify multiple wavelengths simultaneously

Two Types Of Optical Amplifiers

- Erbium doped fiber amplifier (EDFA)
- Raman amplifier appears to be preferred
- Both types can be used in same link
  - Raman amplifier acts as pre-amplifier for EDFA

Erbium Doped Fiber Amplifier

- EDFA
- Original
- Signal amplified in erbium doped fiber
- Fiber stimulated with one or more pump lasers of wavelength different from transmission wavelengths
  - 980 nm, 1480 nm
- Amplifier produces noise: amplified spontaneous emission [ASE]
- Cascaded EDFAs produce increasing noise, which must be filtered out

Erbium Doped Fiber Amplifier 2

- Amplifies multiple wavelengths simultaneously within a range
- Amplification is not uniform with wavelength
- Filters as attenuators used produce equal power in all wavelengths
- Multiple EDFAs used to amplify wide wavelength range
- Used at either beginning or end of link

EDFA

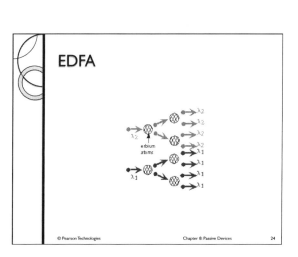

Raman Amplification

- Stimulated Raman scattering is mechanism of operation [SRS]
- Transmission fiber is the amplifying medium
- Used at end of link
- Pump wavelength ~1060 nm
- Amplification occurs along transmission fiber
- Pump power launched from output end towards input end

Stimulated Raman Scattering

- Raman scattering is inelastic scattering of photons
- Raman scattering results in photon scattering and creating a new photon at a a wavelength already present in the fiber

End Of Chapter 8

Questions?
Comments?
Observations?

Fiber Cable Testing, Certification And Troubleshooting

Developed And Delivered By
Eric R. Pearson, CFOS
FOA Master Instructor
BICSI Master Instructor
Pearson Technologies Inc.

Optoelectronics
Chapter 9

Without these, the links are wasted!

Objective And Scope

- In this chapter, you will learn the
  - Definition
  - Function
  - Active devices
  - Performance
  - Configurations
  - OPBR calculation
  - Excess power calculation

Benefits??

- With this knowledge, you will be able
  - Specify requirements for optoelectronics

Essential Optoelectronics Information

- Definition
- Function
- Active devices
- Performance
- Configurations
- OPBR calculation
- Excess power calculation
- OPBR statement

Definition

- Optoelectronics are the devices on the end of the link
- The term 'optoelectronics' describes both the transmitter and receiver, since both function with electrical and optical signals

Fiber Cable Testing, Certification, And Troubleshooting Chapter 9: Optoelectronics

Reminder

- Fiber optic data communications links are duplex

Duplex Communication

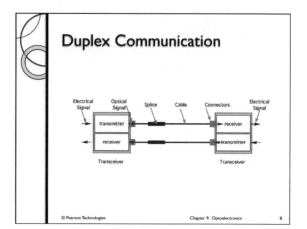

Duplex Operation On 1 Fiber

- ?? figure?? With coupler/splitter

Essential Optoelectronics Information

- Definition
- Function
- Active devices
- Performance
- Configurations
- OPBR calculation
- Excess power calculation
- OPBR statement

Function= Conversion

- The transmitter converts an electrical signal to an optical signal; the receiver performs the reverse conversion
- Conversion is done in the 'active device'
- The key design concern is accuracy of signal conversion
- The optoelectronics achieve high accuracy when
  - Pulse dispersion is sufficiently low
  - Power level at the receiver is proper

Essential Optoelectronics Information

- Definition
- Function
- Active devices
- Performance
- Configurations
- OPBR calculation
- Excess power calculation

Fiber Cable Testing, Certification, And Troubleshooting — Chapter 9: Optoelectronics

Three Transmitter Types

- Light emitting diode (LED)
- Vertical cavity, surface emitting laser (VCSEL)
  - 1 Gbps
  - 10 Gbps
- Laser diode (LD)

LED Characteristics

- Large spot size (150-250 µm)
- Circular spot shape
- Relatively large angle of divergence (0.20-0.25 NA)
- Relatively low bit rate/ bandwidth capability (< 200 Mbps)
- Relatively low launch power (1-10 µW)
- Relatively low cost
- Multimode operation at 850 nm and 1300 nm

Use

- Low bit rate ~200 Mbps
- Short distance < 2 km
- Multimode

LD

- Small spot size
- Rectangular spot shape (2µ ×10µ)
- Small angle of divergence
- Relatively high bit rate/ bandwidth capability (~25 Gbps)
  - Some at 40 Gbps
- Relatively high launch power (>1 mW)
- Relatively high cost
- Singlemode LDs operate between 1310 nm and 1550 nm

Use

- High bit rate
- Long distance
- Singlemode links

VCSEL

- Medium spot size
- Annular spot shape
- Small angle of divergence
- Relatively high bit rate
- Relatively high launch power
- Relatively low cost
- 850 nm multimode VCSELs common
- Available in arrays

USE

- Short distance < 1km
- 1-100 Gbps
- Multimode fiber

Unique Characteristic

- Annular spot shape
- Launches very little power in center of fiber
- Shape eliminates splitting of pulse

Annular Launch Area

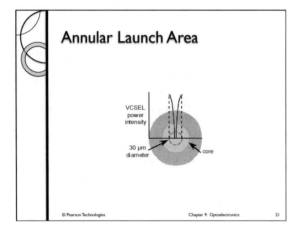

Two Active Devices Dominate

- VCSEL
  - With multimode fiber at 1 and 10 Gbps
  - Arrays for multimode transmission at ≥ 40 and 100 Gbps
- LD for singlemode transmission

Two Receiver Active Devices

- Photodiodes
  - Used in data networks
- Avalanche photodiodes
  - Used in telephone and CATV networks

Essential Optoelectronics Information

- Definition
- Function
- Active devices
- Performance
- Configurations
- OPBR calculation
- Excess power calculation
- OPBR statement

Fiber Cable Testing, Certification, And Troubleshooting   Chapter 9: Optoelectronics

Performance Requirements

- Optical power budget available (OPBA)
- Minimum required loss
- Wavelength
- Spectral width
- Optical connector style
- Electrical signal connector type
- Power requirements
- Environmental requirements
- Signal accuracy

Optical Performance

- OPBA
- Data rate
- RML
- Wavelength
- Spectral width

The Good News

- The data standards define most of the specifications for the optoelectronics

Data Standards Define

- Data rate
- Wavelength
- Spectral width
- Type of source
- Electrical interface
- Bit error rate

Dispersion Controlled By

- Transmitter source
  - Wavelength
  - Spectral width
- Type of source
- Core diameter
- NA
- Distance of transmission

Designer Controls Only

- Distance of transmission
  - Defined by map (Chapter 3)

Bottom Line

- Dispersion Not An Issue
  - …as long as transmission distance is less than distance supported by data standard

OPBA (11-2)

- Is the maximum loss that can occur between a transmitter and receiver
- Value depends on standard

Channel Loss For OM-2

	1000BASE-SX	10GBASE-SX
Core	50	50
BWDP	500/OM-2	500/OM-2
Wavelength	850	850
Channel loss	3.56	1.8
Distance, m	550	82

Channel Loss For OM-3

	1000BASE-SX	10GBASE-SX
Core	50	50
BWDP	1500/OM-3	1500/OM-3
Wavelength	850	850
Channel loss		2.6
Distance, m	550	300

Channel Loss For OM-4

	1000BASE-SX	10GBASE-SX
Core	50	50
BWDP	3500/OM-4	3500/OM-4
Wavelength	850	850
Channel loss		2.6
Distance, m	1000	550

40 Gbps Channel Loss

	40GBASE-SX	40GBASE-SX
Core	50	50
BWDP	3500/OM-3	3500/OM-4
Wavelength	850	850
Channel loss	1.9	1.5
Distance, m	100	150

Fiber Cable Testing, Certification, And Troubleshooting   Chapter 9: Optoelectronics

100 Gbps Channel Loss

	100GBASE-SX	100GBASE-SX
Core	50	50
BWDP	3500/OM-3	3500/OM-4
Wavelength	850	850
Channel loss	1.9	1.5
Distance, m	100	150

Channel Loss?

- Note that the term in previous tables is 'channel loss'
- This value is calculated from the cable attenuation rate and connector loss
- It is not the same as OPBA
- The Gigabit Ethernet standard specifies an OPBA of 7.5 dB
- However, this value is reduced to compensate for dispersion
- If dispersion is not the limiting factor, the maximum link loss can be 7.5 dB

OPBAs And Channel Losses

Standard	Core diameter, µ	Wavelength, nm	OPBA, dB
100BASE-FX	62.5	1300	11
100BASE-SX	62.5	850	4.0
1000BASE-SX	62.5	850	2.33
1000BASE-SX	62.5	850	2.53
1000BASE-SX	50	850	3.25
1000BASE-SX	50	850	3.43
1000BASE-LX	62.5	1300	2.32
1000BASE-LX	50	1300	2.32

10GBASE-SX OPBA And Channel Losses

Core, µ	62.5	62.5	50	50	50	
Modal bandwidth	160	200	400	500	2000	MHz-km
Link budget	7.3	7.3	7.3	7.3	7.3	dB
Distance	26	33	66	82	300	m
Link loss	1.6	1.6	1.7	1.8	2.6	dB
Power penalty	4.7	4.8	5.1	5.0	4.7	dB

Singlemode 10GBASE-LX

Link power budget	9.4	dB
Transmission distance	10	km
Link loss	6.2	dB
Power penalty	3.2	dB

Required Minimum Loss

- The minimum loss that must occur between a transmitter and receiver to avoid overloading
- For most standards, RML=0 dB
- For extended distance singlemode optoelectronics, RML> 0 dB
- Extended distance optoelectronics (-LR, -ER) can overload receiver on short links

Wavelength

- Wavelength is the nominal wavelength of the transmitter
- Wavelength is one of the factors that determines dispersion and power loss in the link

Spectral Width

- Influences dispersion
- Is specified by the data standard
- Designer can specify this with 'must comply with 1000BASE-SX' or equivalent statement

Optical Connector Type

- Optoelectronics are available with many interfaces
- Designer must choose
- Single connector type in network is convenient but not necessary
- If optoelectronics not available in connector type preferred, use patch cords with different connector types on ends

Electrical Connector Type

- Usually determined by standard or by the equipment into which fiber converter (GBIC, etc.) is placed

Essential Optoelectronics Information

- Definition
- Function
- Active devices
- Performance
- Optical Power Requirements
- OPBR calculation
- Excess power calculation
- OPBR statement

Two Receiver Requirements

- Sufficient power
- Not excess power

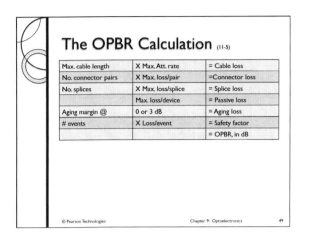
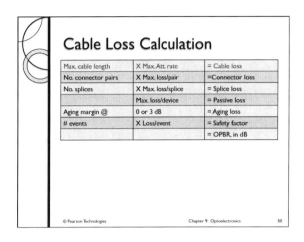
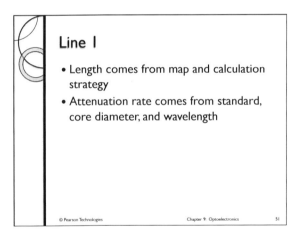
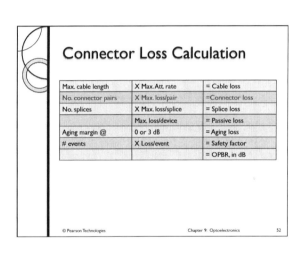
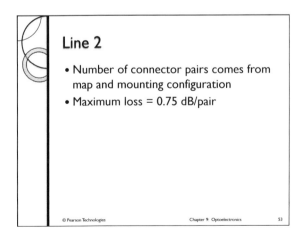
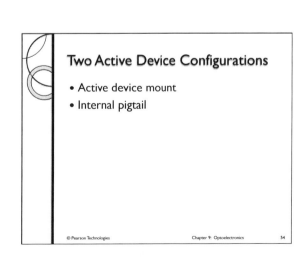

Fiber Cable Testing, Certification, And Troubleshooting   Chapter 9: Optoelectronics

Active Device Mount

- Fiber connects directly to active devices (light source and detector)
- Consequence: end plugs do not count as source of loss
  ◦ Remember connectors are rated dB/pair?
- Multimode optoelectronics have active device mounts

Number of Connector Pairs

- Is not number of plugs divided by two
- Is number of plugs divided by two less one pair
- Connector pair loss, in dB/pair, requires moving light from fiber to fiber
- Moving light from source to fiber or fiber to detector requires one plug, not a pair

Active Device Mount

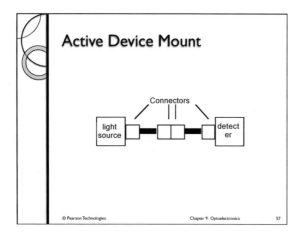

Internal Pigtail

- Fiber connects to connector on back side of front panel with internal pigtail routing light to active devices
- Consequence: end plugs may count as two pairs
  ◦ Remember connectors are rated dB/pair
- Singlemode optoelectronics have internal pigtail

Internal Pigtails

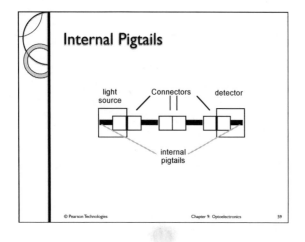

Fiber Mathematics

- 4 plugs/2= 1 pair with ADM
  ◦ Multimode
- 4 plugs/2= 3 pairs with internal pigtail
  ◦ Singlemode
- 4 plugs/2≠ 2 pairs; always!

☺

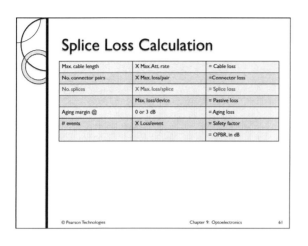

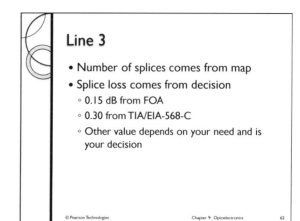

Line 3
- Number of splices comes from map
- Splice loss comes from decision
  - 0.15 dB from FOA
  - 0.30 from TIA/EIA-568-C
  - Other value depends on your need and is your decision

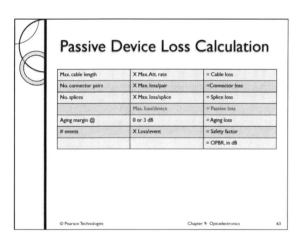

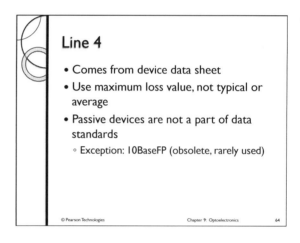

Line 4
- Comes from device data sheet
- Use maximum loss value, not typical or average
- Passive devices are not a part of data standards
  - Exception: 10BaseFP (obsolete, rarely used)

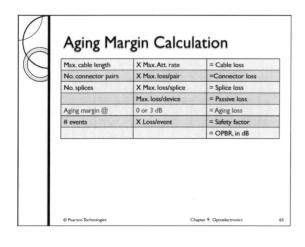

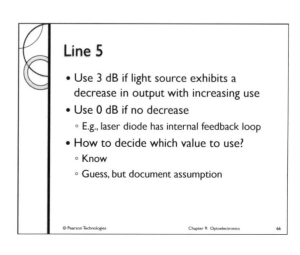

Line 5
- Use 3 dB if light source exhibits a decrease in output with increasing use
- Use 0 dB if no decrease
  - E.g., laser diode has internal feedback loop
- How to decide which value to use?
  - Know
  - Guess, but document assumption

Safety Factor Calculation

Max. cable length	X Max. Att. rate	= Cable loss
No. connector pairs	X Max. loss/pair	= Connector loss
No. splices	X Max. loss/splice	= Splice loss
	Max. loss/device	= Passive loss
Aging margin @	0 or 3 dB	= Aging loss
# events	X Loss/event	= Safety factor
		= OPBR, in dB

Line 6

- From knowledge of environment, you should be able to estimate the number of occurrences of breakage for expected life cycle of network
- If splicing is the repair method, two splices will be assumed per occurrence
- Safety factor is the number of occurrences times loss/occurrence

OPBR Calculation

- This is the calculation of the optical power budget requirement (OPBR)
- The optoelectronics provide the optical power budget available (OPBA)

Summary

OPBR (path) ≤ OPBA (optoelectronics)

Preliminary OPBR Statement

Characteristic	Units	Values
OPBR	dB	≥12, ≥16
Wavelength	nm	850, 1300
Aging margin	dB	0 or 3
Core diameter	μm	62.5 or 50
NA	-	0.275 or 0.20
Mount	-	ADM or internal pigtail
Protocol	--	100BASE-SX
Bit rate	Mbps	100

Activity: OPBR Calculation

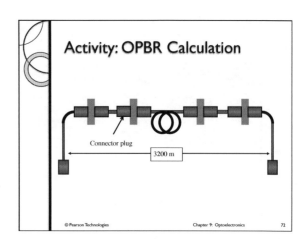

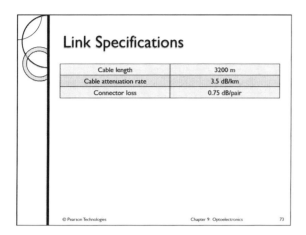

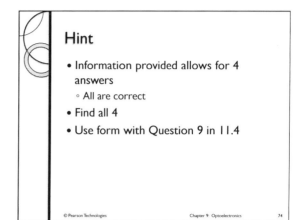

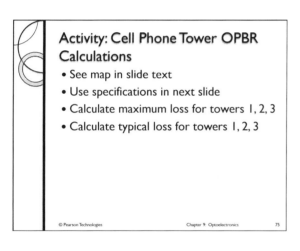

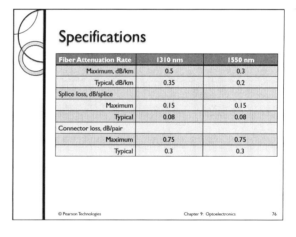

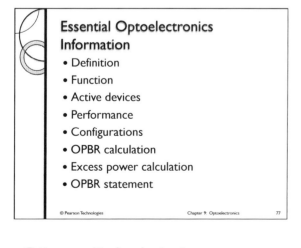

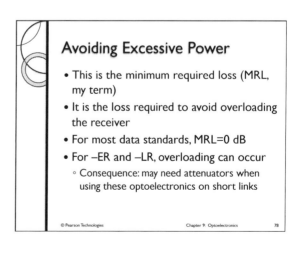

Fiber Cable Testing, Certification, And Troubleshooting — Chapter 9: Optoelectronics

Hooray!!

- We cannot overload receiver!
- As long as we use fully compliant optoelectronics
- But 'extended distance,' singlemode optoelectronics exist
  - They are not fully compliant
  - They achieve extended distance by increasing the power from the transmitter to a level higher than that allowed by the standard

Extended Distance Optoelectronics

- Can overload receiver
  - Whoops!
- To be certain we avoid this potential problem
  - We must perform a MRL calculation
- Note: this calculation helps avoid overloading but does not guarantee no overloading

Excess Power Calculation

Min. cable length	X Typ. Att. rate	= Cable loss
No. connector pairs	X Typ. loss/pair	= Connector loss
No. splices	X Typ. loss/splice	= Splice loss
	Typ. loss/device	= Passive loss
		= MRL, in dB

Excess Power Calculation

Min. cable length	X Typ. Att. rate	= Cable loss
No. connector pairs	X Typ. loss/pair	= Connector loss
No. splices	X Typ. loss/splice	= Splice loss
	Typ. loss/device	= Passive loss
		= MRL, in dB

Line 1

- Use minimum distance in network or minimum distance in group
- Use typical attenuation rate
- Note that this approach does not guarantee no overloading, but it is the best we can do

Excess Power Calculation

Min. cable length	X Typ. Att. rate	= Cable loss
No. connector pairs	X Typ. loss/pair	= Connector loss
No. splices	X Typ. loss/splice	= Splice loss
	Typ. loss/device	= Passive loss
		= MRL, in dB

Fiber Cable Testing, Certification, And Troubleshooting   Chapter 9: Optoelectronics

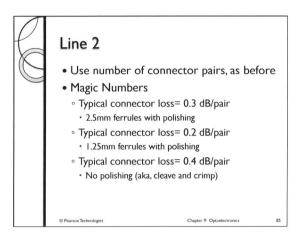

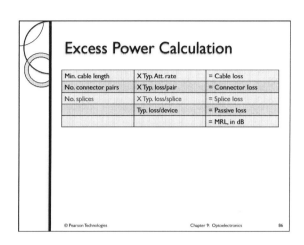

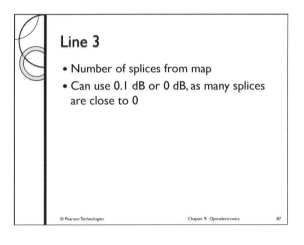

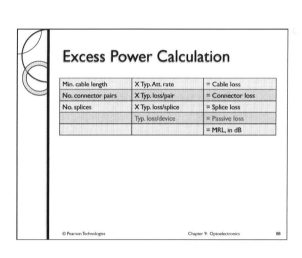

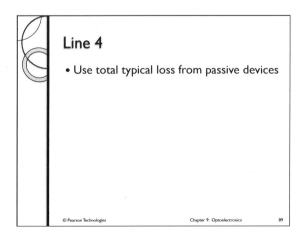

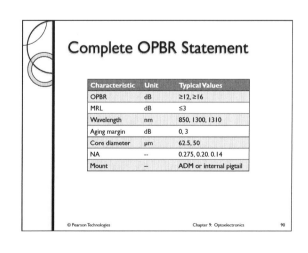

To Be Specified

- Power requirement (110 VAC, 220 VAC, 48 VDC)
- Environment requirements Bandwidth
  ○ Shock, Vibration, operating temperature range
- Bells and whistles
  ○ Features and function in excess of pure transmission function

End Of Chapter 9

Questions?
Comments?
Observations?

Exercises 2, Section 8.6

- Enter
  ○ Characteristic
  ○ Units of measure
  ○ Type of value: nominal, typical, maximum
  ○ Value(s): the values your are likely to specify
- For reason to require specification, use
  ○ PTR- power to the receiver
  ○ ACC- accuracy
  ○ REL- reliability
  ○ FIT- fit
  ○ LGL- legal requirement
  ○ Other

Review Questions

1. What are common design wavelengths?
2. What is a typical optical power budget available from a transmitter-receiver pair?
3. What happens to the OPBA as the bit rate increases?
4. Explain you answer to Question 3.
5. What are common data communication operational wavelengths?

Review Questions

6. Can you explain the reason that a connector at an ADM active device has 'no' loss?
7. What performance characteristics of the optoelectronics are determined by the standard?

End Of Chapter

Questions?
Comments?
Observations?

Cell Phone Tower OPB Calculations

	tower 1	tower 2	tower 3
segment lengths			

(km)

MAXIMUM LOSS

OPBR	tower 1	tower 2	tower 3
1310 nm	dB/km	dB/km	dB/km
cable	dB	dB	dB
connectors			
no. pairs			
connectors	dB/pair	dB/pair	dB/pair
connector loss	dB	dB	dB
splices			
loss/splice	dB	dB	dB
no. splices			
splice loss	dB	dB	dB
cable	dB	dB	dB
connectors	dB	dB	dB
splices	dB	dB	dB
aging margin	dB	dB	dB
TOTAL LOSS, OPBR	dB	dB	dB

Cell Phone Tower OPB Calculations

MAXIMUM LOSS

	tower 1		tower 2		tower 3	
1550 nm		dB/km		dB/km		dB/km
cable		dB		dB		dB
connectors						
connectors		no. pairs		no. pairs		no. pairs
connectors		dB/pair		dB/pair		dB/pair
connector loss		dB		dB		dB
splices						
loss/splice		loss/splice		loss/splice		loss/splice
no. splices				no. splices		no. splices
splice loss		dB		dB		dB
cable		dB		dB		dB
connectors		dB		dB		dB
splices		dB		dB		dB
aging margin		dB		dB		dB
		dB		dB		dB

Cell Phone Tower OPB Calculations

TYPICAL LOSS							
1310 nm							
		dB/km		dB/km		dB/km	
cable		dB		dB		dB	
connectors							
no. pairs				no. pairs		no. pairs	
dB/pair				dB/pair		dB/pair	
connector loss		dB		dB		dB	
splices							
loss/splice							
no. splices		dB		dB		dB	
splice loss						dB	
		dB		dB			
cable							
connectors		dB		dB		dB	
splices		dB		dB		dB	
aging margin		dB		dB		dB	
		dB		dB		dB	
		dB		dB		dB	
TYPICAL LOSS		dB		dB		dB	

ACCEPTANCE VALUES			
TYPICAL LOSS			
MAXIMUM LOSS			

INSERTION LOSS
ACCEPTANCE VALUE

OTDR
ACCEPTANCE VALUES

attenuation rate
splice loss
connector loss

Cell Phone Tower OPB Calculations

TYPICAL LOSS

1550 nm
cable	___ dB/km	___ dB/km	___ dB/km
	___ dB	___ dB	___ dB

connectors
connectors	___ no. pairs	___ no. pairs	___ no. pairs
connectors	___ dB/pair	___ dB/pair	___ dB/pair
connector loss	___ dB	___ dB	___ dB

splices
loss/splice	___ dB	___ dB	___ dB
no. splices	___	___	___
splice loss	___ dB	___ dB	___ dB

cable	___ dB	___ dB	___ dB
connectors	___ dB	___ dB	___ dB
splices	___ dB	___ dB	___ dB
aging margin	___ dB	___ dB	___ dB
	___ dB	___ dB	___ dB

| ___ dB | ___ dB | ___ dB |

ACCEPTANCE VALUES ___
TYPICAL LOSS ___
MAXIMUM LOSS ___

INSERTION LOSS
ACCEPTANCE VALUE ___   ___   ___

OTDR
ACCEPTANCE VALUES

attenuation rate ___
splice loss ___
connector loss ___

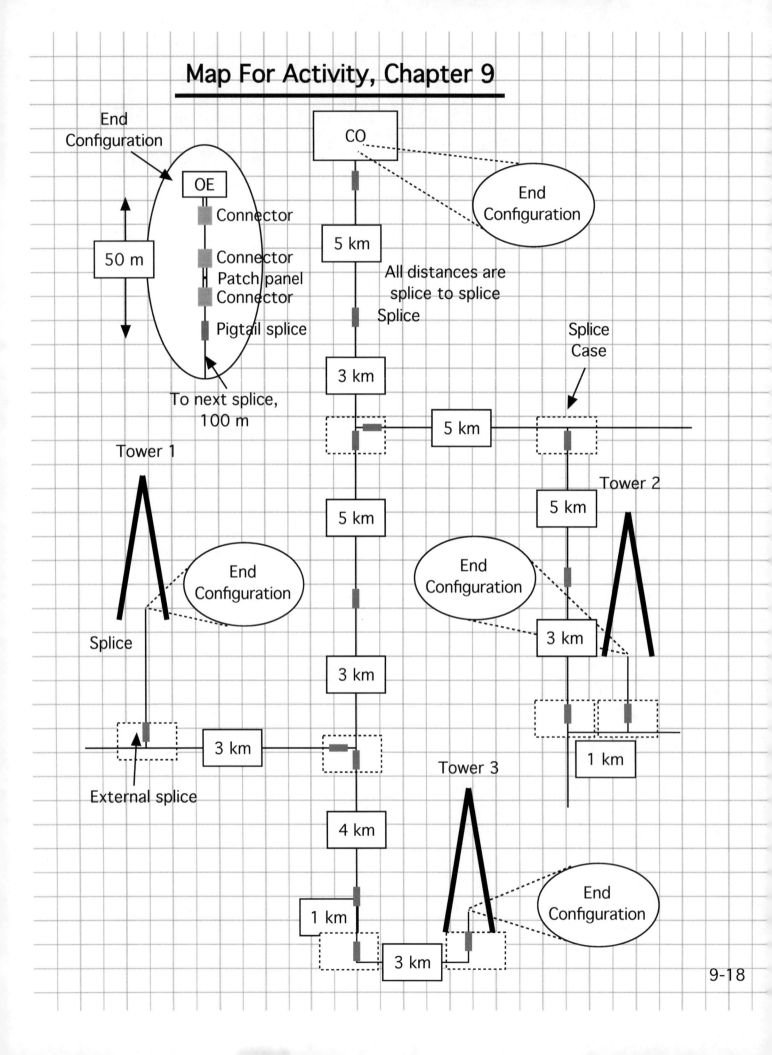

Fiber Cable Testing, Certification, And Troubleshooting — Chapter 10: Insertion Loss Testing

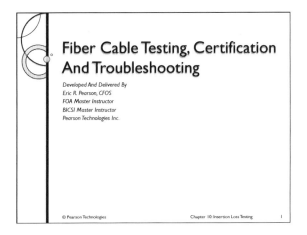

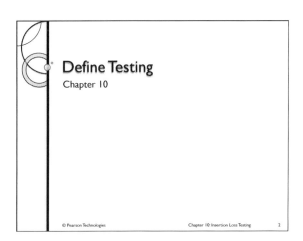

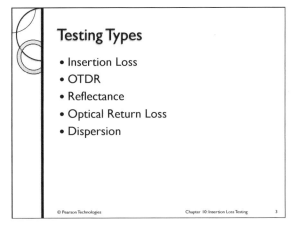

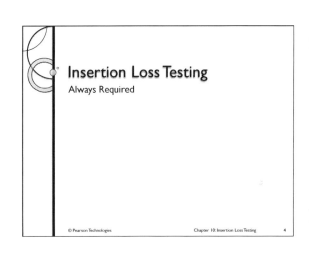

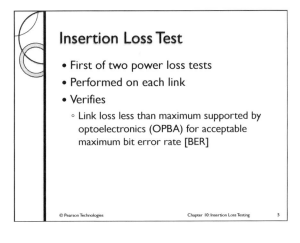

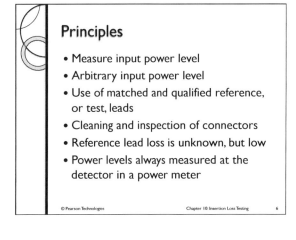

Fiber Cable Testing, Certification, And Troubleshooting   Chapter 10: Insertion Loss Testing

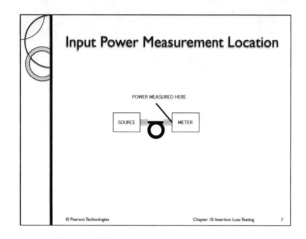

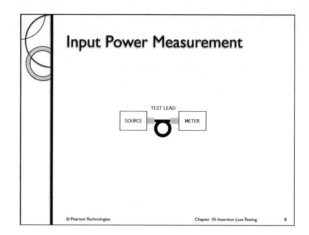

Input Power Arbitrary

- Loss measurements are the same with different input power levels

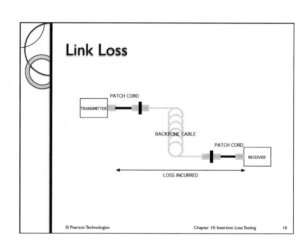

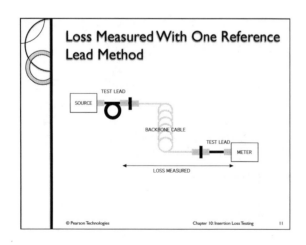

1 Lead Test Method Consequence

- Uses test leads to simulate loss of patch cords on end of backbone
- Measures input power with no connector pairs
- Measures output power with at least two connector pairs
- 2 connectors/2≠1 pair
  ◦ 2 connectors/2=2 pairs
  ◦ in Method B testing!

Fiber Cable Testing, Certification, And Troubleshooting   Chapter 10: Insertion Loss Testing

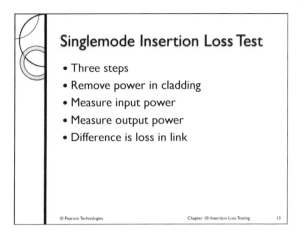

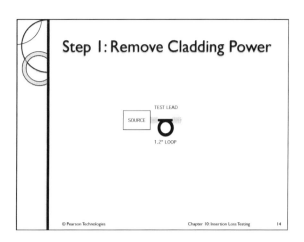

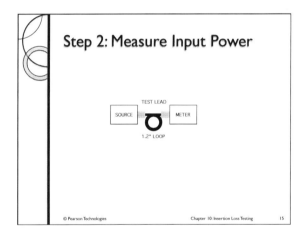

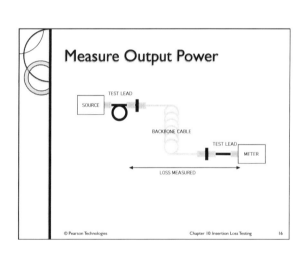

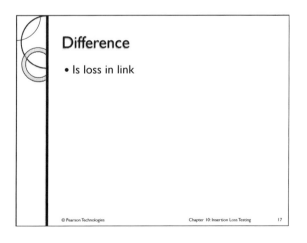

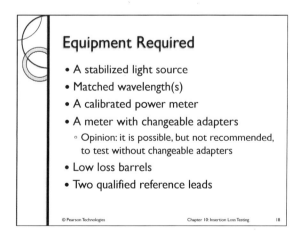

Light Source

- Stabilized = constant output power
  - Make many measurements without checking input power level
- Wavelength is same as that of transmitter

Power Meter

- Calibrated and traceable to NIST
- Insertion loss Method A.1 requires changeable adapters for singlemode testing

Power Meter

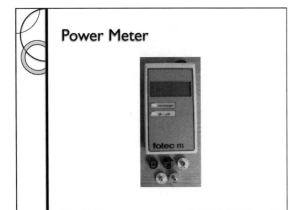

Low Loss Barrels

Matched Reference Leads

- The same core diameter or mode field diameter
- The same NA
- The same connector type
- 1-5 m long
- Leads qualified
  - Qualified means ≤ 0.5 dB/pair (Opinion)
  - TIA/EIA-568-C allows reference leads to be ≤ 0.75 dB/pair

Do Reference Leads Effect Measurement?

- You might argue that the loss of the reference lead will influence the measured loss
- You would be correct
- Prior to use, reference leads are qualified
- Qualified means low loss
  - Low loss is either 0.75 dB or 0.50 dB
- Measurements with different reference leads will differ by ~0.1-0.2 dB

Singlemode Reference Lead Qualification

- Qualified means low loss
- Low loss can means two values
  - 0.75 dB/pair (TIA/EIA-568-C)
  - 0.50 dB/ pair (recommended)
- Input reference lead has 1.2" diameter loop
  - 2μ by 10 μ LD against 8.2μ core puts optical power in cladding
  - Loop removes power from cladding

First Two Qualification Steps

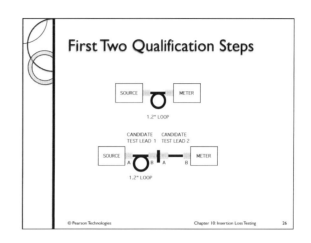

Third Qualification Test

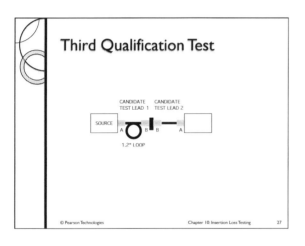

Difference Is Loss

- Difference between power level with one reference lead and that with two leads is loss of connector pair
- Determine whether this loss is acceptable

Recommendation (Opinion)

- Use as acceptance value for qualification
  - 0.5 dB instead of 0.75 dB as allowed by TIA/EIA-568-C
- Test leads should have loss lower than cables being tested
- Risk of using 0.75 dB: acceptance of high loss products by using reference leads that are 'high' loss

Summary: Singlemode Testing

- Source reference
  - Single lead method
  - 1.2" diameter loop in source test lead
- Test source
  - Stabilized
  - Wavelength matches that of transmitter
- Power meter
  - Traceable calibration
  - Changeable adapters
- Test leads are
  - Matched
  - Qualified
  - 1-5 m long
- Barrels are low loss

Fiber Cable Testing, Certification, And Troubleshooting — Chapter 10: Insertion Loss Testing

Multimode Insertion Loss Test

- At least four methods
  - Method B with mandrel
  - Method B with EF-compliant source
  - Method B with EF-compliant test lead
  - Fiber vendor-specific test procedure for BIMM
- Other methods in text
  - Two lead reference
  - Three lead reference

Important Note: BIMM Testing

- Bend insensitive multimode fibers [BIMM] require test method different from that in TIA/EIA-568-C or TIA/EIA-568-B
- Standards have not been issued for such testing
- Fiber manufacturer defines test method

Three Steps

- Create proper power distribution in core
- Measure input power
- Measure output power
- Difference is loss in link

Multimode Testing

- May not be simulation
- May be normalized testing
  - Recommended at this time
  - Complies with TIA/EIA-568-B but not TIA/EIA-568-C
- May be 'encircled flux' testing
  - Is current method in data standards but may change in near future
  - Goal is simulation of power loss with VCSEL

Multimode Testing

- Simulation occurs for some, but not all, multimode testing
- Different multimode sources have significantly different characteristics
- Differences are in
  - NA/angle of divergence
  - Spot size
- These differences result in significantly different insertion loss values

Attenuation Rate Vs. Divergence Angle

- Attenuation rate in core depends on travel path in core
- Longer the path, higher the attenuation rate
- Rays at large angles to core axis travel further than rays at small angles
- LEDs have large angles of divergence
- VCSELs have small angles

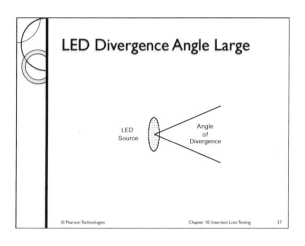

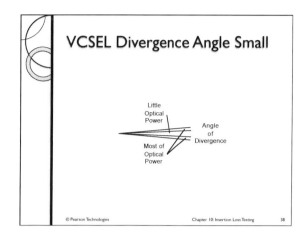

Comparison
- Attenuation rate with LEDs high
- Attenuation rate with VCSELs low

Spot Size Comparison
- LED spot size large ~ 150μ
- VCSEL spot size small ~ 30μ

Connector Loss
- Determined by power at core boundary
- LED large spot size creates much power at boundary
- VCSEL small spot size creates little power at boundary
- Connector loss with LED high
- Connector loss with VCSEL low

Comparison

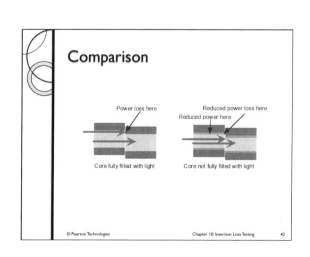

Consequence

- Differences in NA/angle of divergence and spot sizes result in significantly different insertion loss values
- Example: 2.7 vs. 3.7 dB

TIA/EIA-568-B Multimode Method

- Method prior to TIA/EIA-568-C
- Is 'normalized' method
- Method produces same power distribution at end of source lead with different sources
- Power distribution may comply with that required by TIA/EIA-568-C

Normalized Testing

- Replaces multiple tests with single test
- Angular and radial distribution is compromise of those from LEDs and VCSELs
- Presented in TIA/EIA-568-B (2000)

Normalized Test Requirements

- Category 1 source
  - Category defines ratio (CPR) of optical power in center to total in core
  - Overfills both NA and core diameter
- Mandrel on test lead
  - Removes some power
    - At high angle to axis
    - At core boundary
- Required for testing at 850 and 1300 nm
- Mandrel use provides same loss measurements with different sources

Mandrel

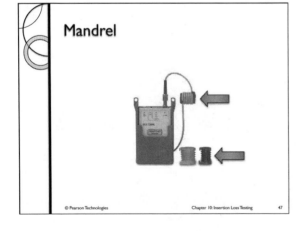

Mandrel Sizes

Core diameter, μm	Mandrel diameter for 900 μm	Mandrel diameter for 3 mm
50	1.0 in.	0.9 in.
62.5	0.8 in.	0.7 in.

Multimode Reference Lead Qualification

- Same as TIA/EIA-568-C method except
  - Use mandrel on multimode reference lead instead of 1.2" loop on singlemode lead
- Opinion: recommended method
  - Do not use mandrel
  - Use 0.5 dB instead of 0.75 dB as acceptance value

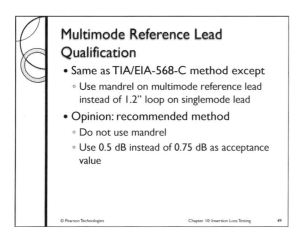

Multimode Test Lead Qualification

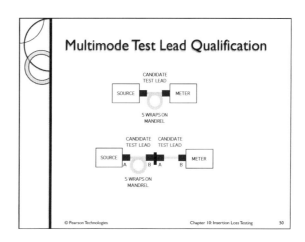

Third Step

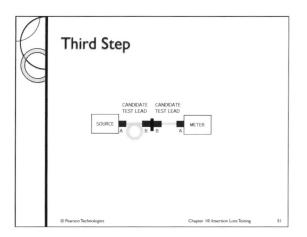

Caution

- This procedure will provide valid test results as long as
  - End A of candidate 1 is always plugged into the source
  - End B of candidate 1 is not damaged or dirty

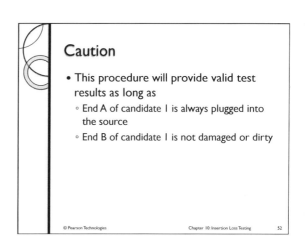

Measure Input Power, Multimode

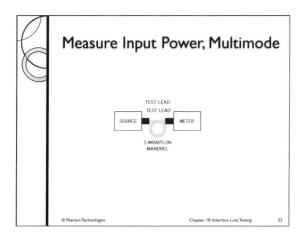

Measure Output Power, Multimode

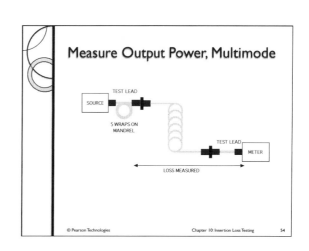

Difference

- Is loss in link

Encircled Flux (EF) Testing

- Normalized testing does not simulate VCSEL sources used from 1-100 Gbps on multimode fibers
- Encircled flux defines power distribution that simulates VCSEL power distribution
  ○ Remember annular launch conditions?
- Required by TIA/EIA-568-C for multimode fibers
- Used for 50μ only, not for 62.5μ

EF Launch Condition

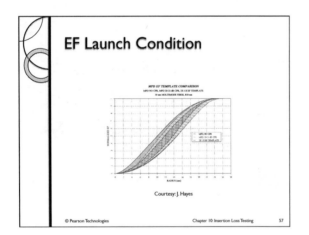

Courtesy: J. Hayes

Measure Input Power With EF Source

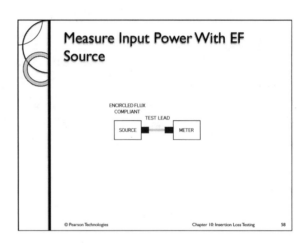

Measure Output Power, Multimode

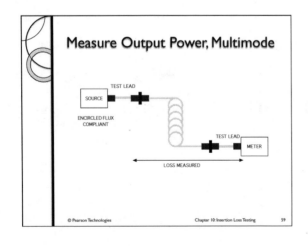

Difference

- Is loss in link

Fiber Cable Testing, Certification, And Troubleshooting   Chapter 10: Insertion Loss Testing

Input Power Measurement With EF Compliant Test Lead

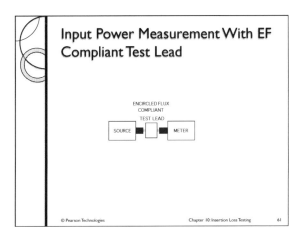

EF Loss Measurement

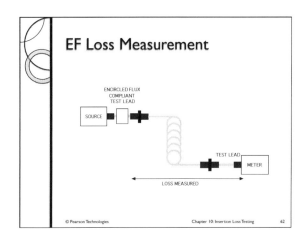

Three EF Problems

- Only one manufacturer of EF compliant source [Kingfisher, Australia]
- Only one manufacturer of EF compliant launch cable used instead of source [England]
- Launch cable expensive and will need replacement or repair when damaged
  - ~ $250 x 2 [working and spare]

Recommendation/Opinion

- This author recommends the use of the mandrel wrap for multimode testing
- This method provides results that are consistent when different light sources are used
- Wait for next revision to test standard

Summary: Multimode Testing

- Reference
  - Is single lead input power measurement
  - Has mandrel sized to core and jacket diameters
- Test source
  - Stabilized
  - Wavelength matches that of transmitter
  - Category 1 source- recommended
    - EF launch conditions in future [opinion]
- Power meter
  - Traceable calibration
  - Changeable adapters
- Test leads are
  - Matched
  - Qualified
  - 5 m long
- Barrels are low loss

Directional Differences

- Would you expect that the insertion loss is *exactly* the same in both directions?
- Of course not

Fiber Cable Testing, Certification, And Troubleshooting   Chapter 10: Insertion Loss Testing

Causes Of Directional Differences

- Core diameter
- Core offset
- NA
- Clad diameter
- Cladding non circularity
- Offset of fiber in ferrule
- Differential modal attenuation (multimode only)
  - Multimode connectors experience decreasing loss as distance from the transmitter increases

Test In Both Directions?

- No
- Typical directional differences are small, on the order of 0.2 dB
- TIA/EIA-568-C allows testing in one direction
- Many telephone companies require testing in both directions to ensure lack of problems

Range Testing

- Would you expect every measurement of same link to be exactly the same?

Range

- Why not?
- Answer

Determine Range

- Make 6 measurements of insertion loss on 6 fibers in network
- Calculate difference between maximum and minimum loss on each fiber
- Now have 6 range values
- Choose maximum range as range for network
- Any future increase in insertion loss greater than range indicates degradation of link
- Any increase less than range indicates normal behavior

Purpose Of Range Testing

- To simplify interpretation of increased loss during troubleshooting and maintenance activities
- Without a range value, how do you determine whether increase loss is normal behavior or due to degradation of link?

Fiber Cable Testing, Certification, And Troubleshooting Chapter 10: Insertion Loss Testing

Insertion Loss Test Advantages

- Relatively low cost test equipment
- Easy test to perform
- Simple test to interpret

Insertion Loss Test Disadvantages

- Test is blind to location and distribution of loss
- Conditions of reduced reliability can result in acceptable insertion loss values
- In other words, high loss in part of link can be concealed by low loss in part of link

Have Acceptable Loss From…

- Typical loss connectors and typical attenuation rate cable
- Low loss connectors and unacceptably high attenuation rate cable
- Unacceptably high loss connector(s) and low attenuation rate cable
- Low loss connectors, low attenuation rate cable, and a bend radius violation, or violation of some other cable performance parameter(s)

Conditions of Reduced Reliability

- Three of the 4 conditions in the previous slide represent conditions of reduced reliability

Additional Test Needed

- To indicate that all components in a link are properly and reliably installed
- Components are
  - Connectors
  - Cables
  - Splices

We need an additional test!

Fiber Cable Testing, Certification, And Troubleshooting — Chapter 10: Insertion Loss Testing

That Test Is…

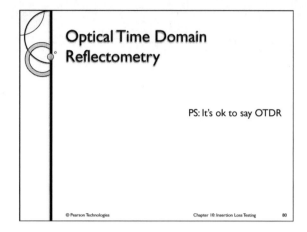

Optical Time Domain Reflectometry

PS: It's ok to say OTDR

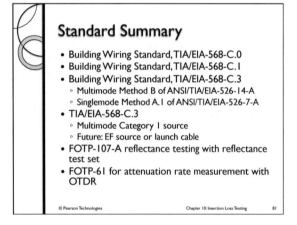

Standard Summary

- Building Wiring Standard, TIA/EIA-568-C.0
- Building Wiring Standard, TIA/EIA-568-C.1
- Building Wiring Standard, TIA/EIA-568-C.3
  - Multimode Method B of ANSI/TIA/EIA-526-14-A
  - Singlemode Method A.1 of ANSI/TIA/EIA-526-7-A
- TIA/EIA-568-C.3
  - Multimode Category 1 source
  - Future: EF source or launch cable
- FOTP-107-A reflectance testing with reflectance test set
- FOTP-61 for attenuation rate measurement with OTDR

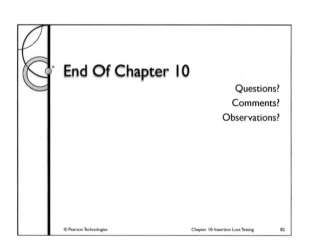

End Of Chapter 10

Questions?
Comments?
Observations?

Singlemode Insertion Loss Measurement Data Form

USE SINGLEMODE INSERTION LOSS TEST SET, ONE LEAD REFERENCE

UNICAM 1

Direction	1310 nm	1550 nm
Source at End A		
Source at End B		

UNICAM 2

Direction	1310 nm	1550 nm
Source at End A		
Source at End B		

USE SINGLEMODE INSERTION LOSS TEST SET, ONE LEAD REFERENCE

LINK # A

Direction	1310 nm	1550 nm
Source at reel 1		
Source at reel 4		

LINK # B

Direction	1310 nm	1550 nm
Source at reel 1		
Source at reel 4		

LINK # C

Direction	1310 nm	1550 nm
Source at reel 1		
Source at reel 4		

LINK # D

Direction		
Source at re		
Source at re		

USE OTDR WITH LAUNCH CABLE AT BOTH ENDS; SET CURSARS TO SIMULATE INSERTION LOSS

Direction	1310 nm	1550 nm
Source at reel 1		
Source at reel 4		

Direction	1310 nm	1550 nm
Source at reel 1		
Source at reel 4		

Direction	1310 nm	1550 nm
Source at reel 1		
Source at reel 4		

Direction		
Source at re		
Source at re		

DIFFERENCES

Direction	1310 nm	1550 nm
Source at reel 1		
Source at reel 4		

Direction	1310 nm	1550 nm
Source at reel 1		
Source at reel 4		

Direction	1310 nm	1550 nm
Source at reel 1		
Source at reel 4		

Direction		
Source at re		
Source at re		

© Pearson Technologies Inc.

Fiber Cable Testing, Certification, And Troubleshooting   Chapter 11: OTDR Testing

Fiber Cable Testing, Certification And Troubleshooting

Developed And Delivered By
Eric R. Pearson, CFOS
FOA Master Instructor
BICSI Master Instructor
Pearson Technologies Inc.

OTDR Testing
Chapter 11

Testing Types

- Insertion Loss
- OTDR
- Reflectance
- Optical Return Loss
- Dispersion

Insertion Loss Test Disadvantages

- Test is blind to location and distribution of loss
- Conditions of reduced reliability can result in acceptable insertion loss values

Acceptable Loss From…

- Typical loss connectors and typical attenuation rate cable
- Low loss connectors and unacceptably high attenuation rate cable
- Unacceptably high loss connector(s) and low attenuation rate cable
- Low loss connectors, low attenuation rate cable, and a bend radius violation, or violation of some other cable performance parameter(s)

Conditions of Reduced Reliability

- Low loss connectors and unacceptably high attenuation rate cable
- Unacceptably high loss connector(s) and low attenuation rate cable
- Low loss connectors, low attenuation rate cable, and a bend radius violation, or violation of some other cable performance parameter(s)

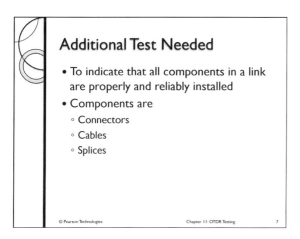
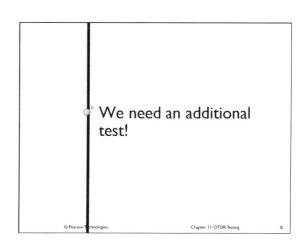
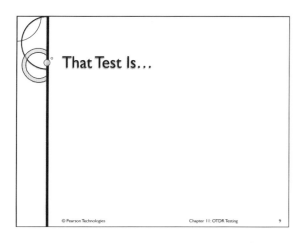
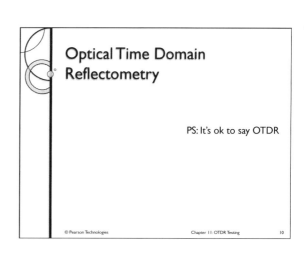
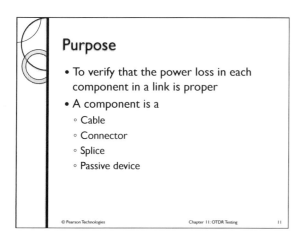
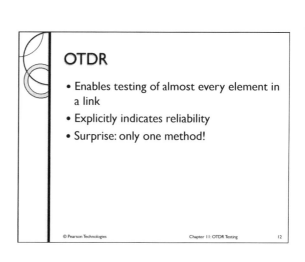

Fiber Cable Testing, Certification, And Troubleshooting — Chapter 11: OTDR Testing

OTDR Principles

- Power loss (attenuation) results from scattering of light by atoms
- Lost power is scattered towards core-cladding boundary outside angle defined by NA
- This mechanism is called Rayleigh scattering (Chapter 3)
- Some lost power is scattered backwards towards boundary inside angle defined by NA
- Some light is always traveling backwards in the fiber

OTDR Measures Backscattered Power

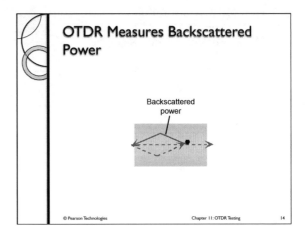

OTDR Functional Diagram

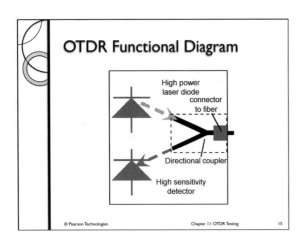

Theoretical OTDR Trace

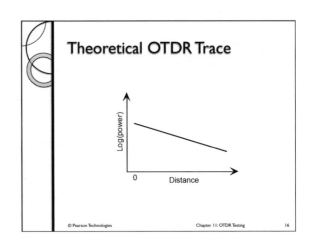

Multiple Measurements

- Backscattered power is small
- Amplification of power in OTDR adds noise
- OTDR makes multiple tests of link, displaying averaged power levels
  - My experience: 1000-8000 tests
- Averaging reduces noise to acceptable level

OTDR Displays Power Differences

- Measure attenuation rate from difference of power levels at beginning and end of segment
- Measure connection loss from difference of power levels before and after connection
- Can measure power loss of almost every component in link

Fiber Cable Testing, Certification, And Troubleshooting — Chapter 11: OTDR Testing

Theoretical OTDR Trace

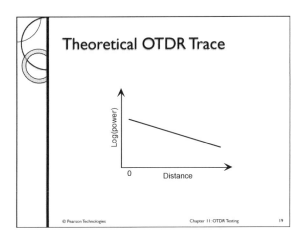

Fiber Ends Create Reflections

- Called Fresnel reflections
- Some connectors and splices create fiber ends
- Power from these reflections adds to the power from Rayleigh scattering

Add Fresnel Reflection Power

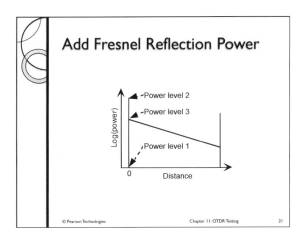

Remember

- Horizontal axis is time
  - Remember the IR?
- The OTDR measures round trip travel time
- The OTDR calculates distance from the IR
- In the previous slide, 0 distance means 0 time

Responding To Power Changes Takes Time!

- The previous figure requires responding to power changes in zero time
- We must modify the previous figure

Response Time Creates Blind/Dead Zones

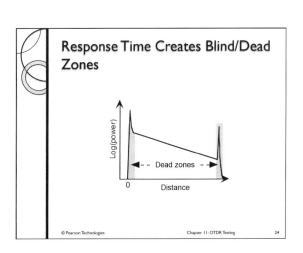

Blind Zones, Dead Zones

- Are areas in which you cannot read the features
- If features (connectors, splices, etc.) are more closely spaced than the width of the dead zone, you will not see the separate features, only the sum of the effects of those features

Concealed Features

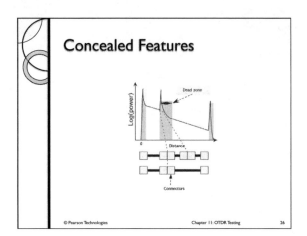

Most Important Rule

- For interpretation of OTDR traces
- You cannot build a map from a trace
  - Dead zones may conceal features!
  - Multiple features can have the same appearance on a trace
- You can only interpret a trace from a map

Three Basic Traces

- Reflective loss
- Non-reflective loss
- Bad launch

Basic Trace #1: Reflective Loss

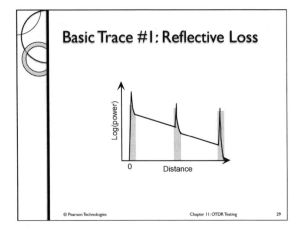

..From At Least Five Configurations

- Two segments connected by radius connectors
- Two multimode segments connected by mechanical splice
- Two singlemode segments connected by a mechanical splice
- A broken fiber and a tight tube cable and
- A single cable segment that is demonstrating multiple reflections, also known as ghost reflections

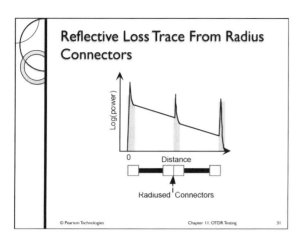

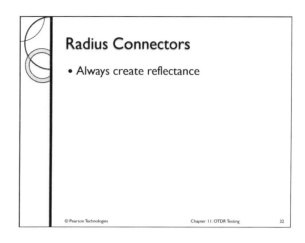

Radius Connectors
- Always create reflectance

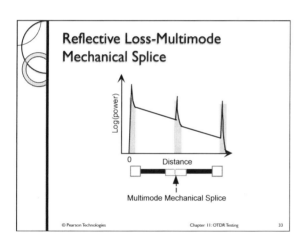

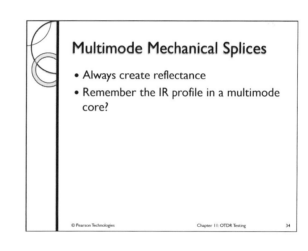

Multimode Mechanical Splices
- Always create reflectance
- Remember the IR profile in a multimode core?

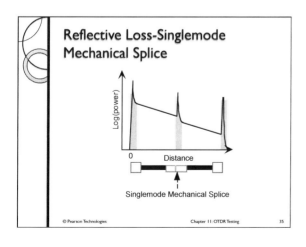

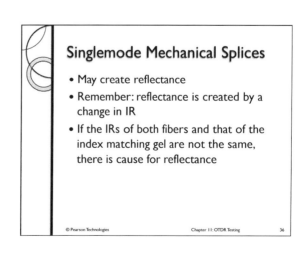

Singlemode Mechanical Splices
- May create reflectance
- Remember: reflectance is created by a change in IR
- If the IRs of both fibers and that of the index matching gel are not the same, there is cause for reflectance

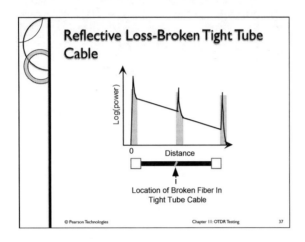
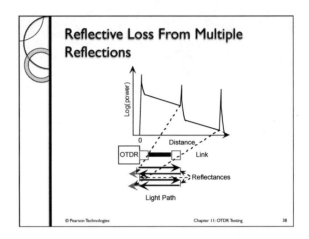
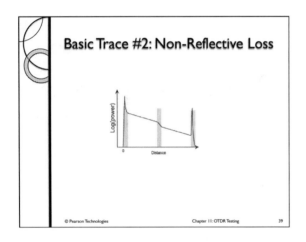

...From At Least 4 Configurations

- Two segments connected with a fusion splice
- Two singlemode segments connected with a mechanical splice
- Two segments connected by APC connectors
- A cable with a violation of any cable performance parameter

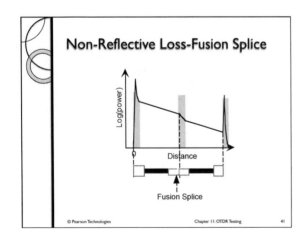
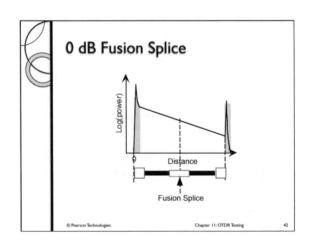

Most Important Rule

- The last slide demonstrates the most important rule for interpretation of OTDR traces
- Most Important Rule
  - You cannot build a map from a trace
  - You can only interpret a trace from a map

Non-Reflective Loss-Singlemode Mechanical Splice

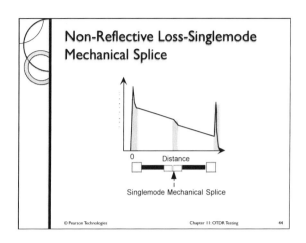

Singlemode Mechanical Splices

- May not create reflectance
- Remember: reflectance is created by a change in IR
- If the IRs of both fibers and that of the index matching gel are the same, there is no cause for reflectance

Non-Reflective Loss-APC Connectors

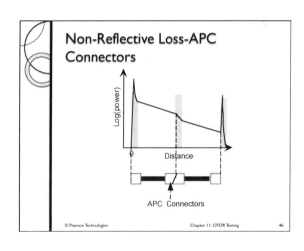

Non-Reflective Loss-Cable Parameter Violation

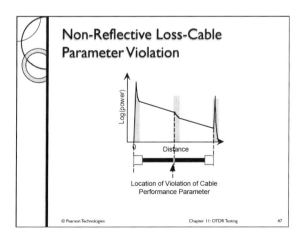

Basic Trace #3: No/Bad Launch

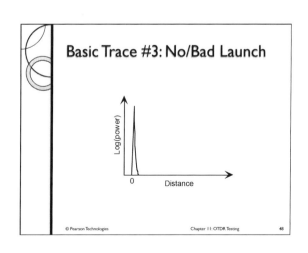

Three Interpretations

1. Broken fiber at OTDR
2. Broken fiber in dead zone
3. Cable shorter than width of dead zone
   ◦ Example

Unusual Trace: No End Reflection

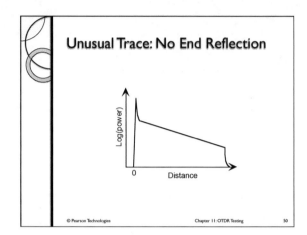

Causes- No End Reflection

- APC connector on end
- Bad/high angle cleave on end
- Mechanical splice on end of singlemode fiber

Using The OTDR
Making Loss Measurements

Data Required For Set Up

- Wavelength of operation
- Pulse width
- Maximum length of cable to be tested
- Index of refraction and
- Maximum time allowed for the test or the number of pulses to be analyzed by the OTDR
   ◦ Determines smoothness of trace=low noise
- Backscatter coefficient
   ◦ Calibrates OTDR to fiber under test

Pulse Width

- Determines power launched into fiber
- High power required for long fibers
   ◦ Use long pulse width for long cables
- High power results in long dead zone
- Low power results in short dead zone
   ◦ Use for short pulse width short cables

Fiber Cable Testing, Certification, And Troubleshooting   Chapter 11: OTDR Testing

Maximum Length Of Cable

- Determines testing time
- Set to value longer than, but close to, length of cable to be tested

Index Of Refraction

- Of fiber under test
- Enables accurate length and attenuation rate measurements

Maximum Time For Test

- Determines accuracy of measurements by reducing noise in trace
- Increased time or number of pulses results in smoother trace and increased accuracy
- Use high enough value to produce smooth trace

Multimode Backscatter Coefficient

Fiber	850 nm	1300 nm
OFS 550/300	-68.4 dB	-75.8 dB
OFS LaserWave G+	-68.4 dB	-75.8 dB
Corning InfiniCor® 62.5	-68 dB	-76 dB
Corning InfiniCor® 50	-68 dB	-76 dB

Singlemode Backscatter Coefficient

Fiber	1310 nm	1550 nm
Corning SMF-28e	-77 dB	-82 dB
Corning Leaf		-81
Draka, G.652	-79.4	-82

Important Fact

- OTDR measures fiber length
- Cable length is less than fiber length

Fiber Cable Testing, Certification, And Troubleshooting — Chapter 11: OTDR Testing

Distance Inaccuracies

Distance, m		100	500	1000	5000	10000
fiber Excess length	buffer tube Excess length			Distance error, m		
0.01%	1.91%	1.92	9.60	19.19	95.97	191.94
0.02%	1.91%	1.93	9.65	19.29	96.47	192.94
0.03%	1.91%	1.94	9.70	19.39	96.97	193.94
0.04%	1.91%	1.95	9.75	19.49	97.47	194.94
0.05%	1.91%	1.96	9.80	19.59	97.97	195.94
0.06%	1.91%	1.97	9.85	19.69	98.47	196.94
0.07%	1.91%	1.98	9.90	19.79	98.97	197.94
0.08%	1.91%	1.99	9.95	19.89	99.47	198.94
0.09%	1.91%	2.00	10.00	19.99	99.97	199.94
0.10%	1.91%	2.01	10.05	20.09	100.47	200.94

Always Use Launch Cable

- To protect the OTDR port
- To measure the loss of the near end connector
  - Launch cable must be longer than dead zone width to measure loss of near end connector
- Launch cable can be longer than longest cable to measure to avoid ghost reflections
  - This may not be practical for singlemode links

OTDR Trace With Launch Cable

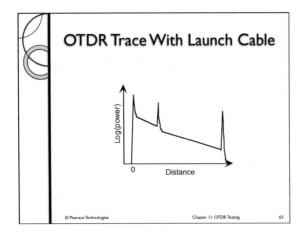

Making Measurements
How To Avoid Cursing At Cursors!

Two Measurement Methods

- Trust (imperfect) software
  - Position cursors to verify proper operation of software
- To be fair, software is pretty good
- Position cursors with knowledge of rules for proper cursor positions

Measurements To Make

- Length
  - First segment
  - Subsequent segments
- Connector and splice loss
  - Estimated method
  - Accurate method
- Attenuation rate

Fiber Cable Testing, Certification, And Troubleshooting — Chapter 11: OTDR Testing

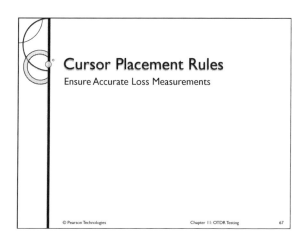

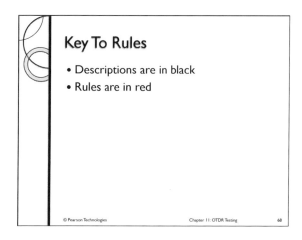

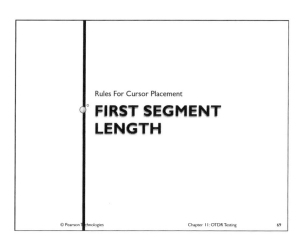

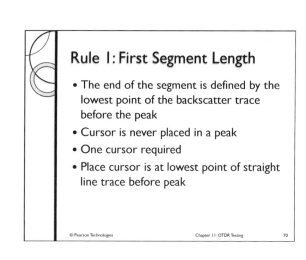

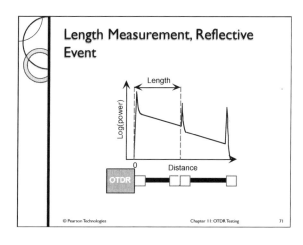

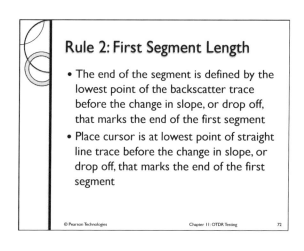

Length Measurement, Non-Reflective Event

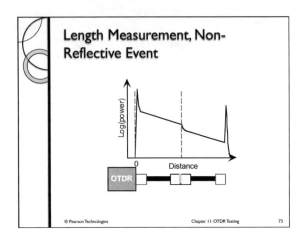

Subsequent Segment Length

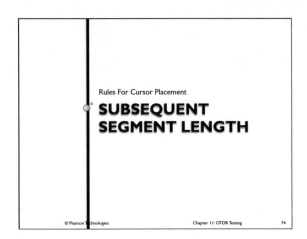

Rules For Cursor Placement

Rule 3: Subsequent Segment Length

- Two cursors required
- Place one cursor is at the lowest point on the straight line trace before the peak that marks the end of the previous segment
- Place the second cursor is at the lowest point of a straight line trace of the segment before either the peak that marks the end of that segment or the non-reflective drop that marks the end of the segment

Length Measurement, Reflective Ends

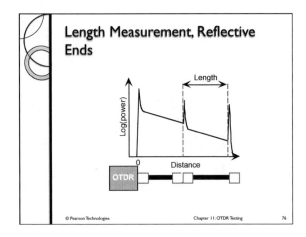

Rule 4: Subsequent Segment Length

- Two cursors required
- Place the first cursor at the lowest point on the straight line trace before the drop that marks the end of the previous segment
- Place the second cursor at the lowest point of a straight-line trace of the segment being measured before the drop that marks the end of that segment

Length Measurement, Non Reflective Ends

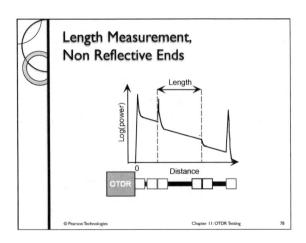

Fiber Cable Testing, Certification, And Troubleshooting — Chapter 11: OTDR Testing

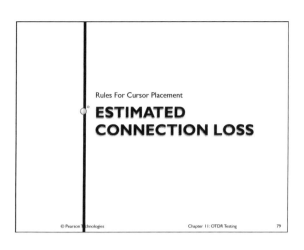

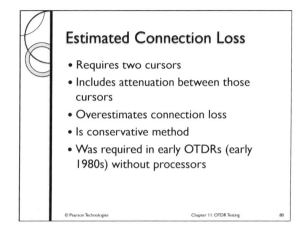

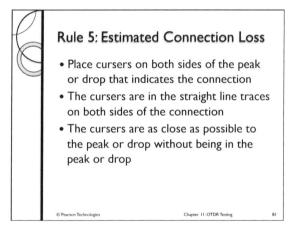

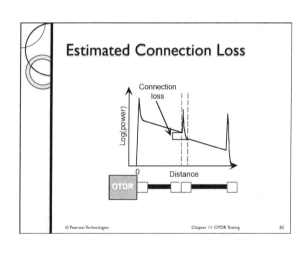

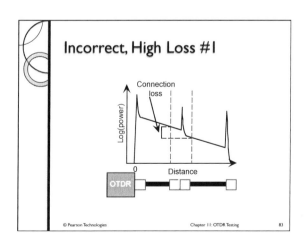

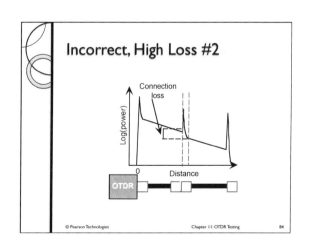

11-14

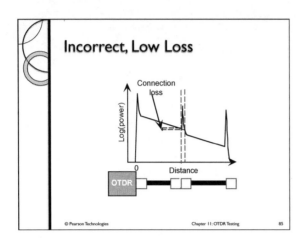
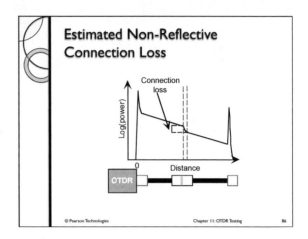

Rule 6: Accurate Connection Loss

- One cursor required
- Computer performs a least squares analysis or best fit of straight-line trace after peak
- Computer calculates loss at connection from extrapolation of trace after peak to trace before peak

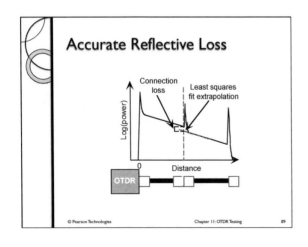
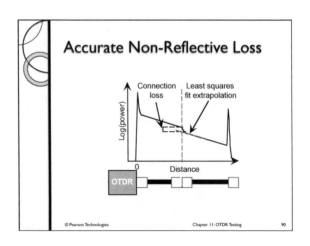

Fiber Cable Testing, Certification, And Troubleshooting — Chapter 11: OTDR Testing

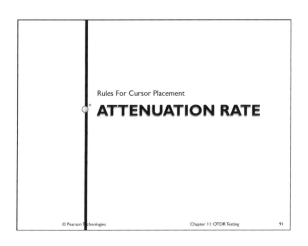

Rules For Cursor Placement
ATTENUATION RATE

Rule 7: Attenuation Rate

- Attenuation rate measurements require two cursors:
  ○ Place cursor at the beginning of the segment and
  ○ Place cursor at the end of the segment
- Place the cursors must be as far apart as possible to create an attenuation rate value that is most representative of the fiber
- The cursors must not be in or enclose any features, such as a drop or a peak

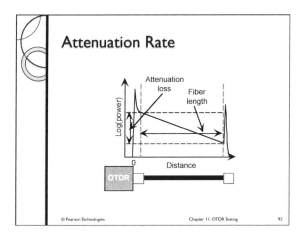

Attenuation Rate

Attenuation Rate Rules

- Place two cursors in same cable segment
- Place cursors in straight line as far apart as possible
- Do not place cursors in peak
- No features can be between cursors
  ○ Features are reflections and drops

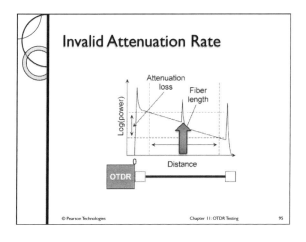

Invalid Attenuation Rate

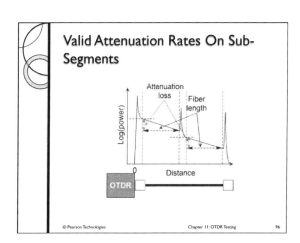

Valid Attenuation Rates On Sub-Segments

Acceptable Placement

- To avoid placing cursors in peak, move cursors closer to each other
- Attenuation rate value will change
- Interpretation will not change!

Alternative Attenuation Rate

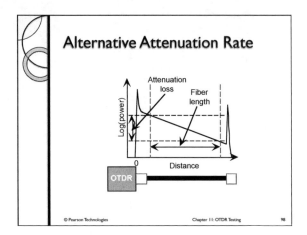

Unusual Traces

- Gainers
  - Not real
  - Have loss in opposite direction
- Multiple reflections, aka 'ghost' reflections
  - Not real
  - Look like broken fibers

Gainers! Amplification? Wonderful!

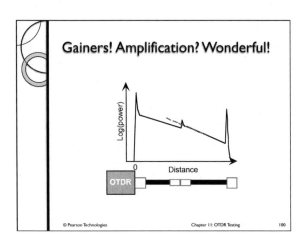

Gainers Not Real

- Result from differences in two fibers
- Differences in
  - Attenuation rate
  - Core diameters or MFDs
- Differences bias OTDR loss measurement
  - Higher in one direction
  - Lower in opposite direction
- A gainer requires loss measurement in opposite direction to prove acceptable loss

Interpreting Gainers

- Multimode gainers
  - Small core to large core
  - Low attenuation rate to high attenuation rate
- Singlemode gainers
  - Large MFD to small MFD

Accurate Splice Loss

- Is the average of the values measured in both directions
- Averaging eliminates the bias

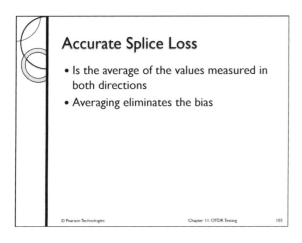

Multiple Reflection

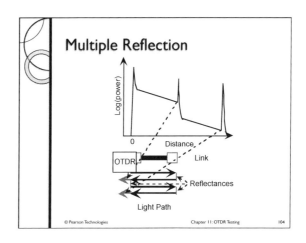

Single Segment With Ghost

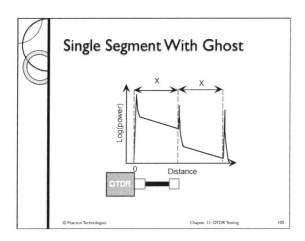

Ghost Problem Significant

- When short segment precedes a long segment
- Ghost from short segment appears in middle of long segment
- Is it a real fiber end/reflection or a ghost?

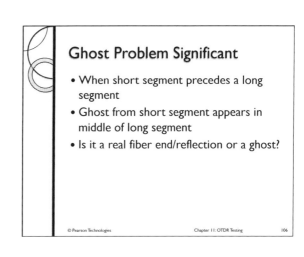

Multiple Segments With Ghost

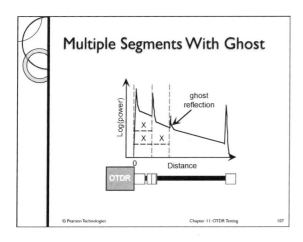

Must Prove Ghost Is Ghost

- Cannot ignore, as a reflectance is a fiber end, unless it is a ghost
- How to prove reflection is ghost?

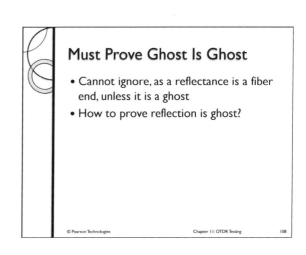

Exorcising The Ghost

- Measure distance to suspected ghost
  - Is should be approximately double the distance of the prior segment
  - If so, it may be a ghost
- Shoot trace from opposite end
  - If feature disappears, it is a ghost
  - If feature stays at same location, it is a ghost
  - If feature moves to appropriate location, it is real

Desperate Measures

- Index matching gel on connectors may reduce reflectance and may eliminate ghost
- In an emergency, baby oil or mineral oil may also reduce reflectance

OTDR Directional Differences

- Caused by differences in
  - Core diameters
  - MFDs
  - NAs
  - Attenuation rates

Define Test Requirements

- Specify insertion loss
  - Method
  - Number of wavelengths
  - Number of directions (1 is acceptable)
  - Acceptance values
- OTDR
  - Number of wavelengths
  - Number of directions (2 necessary for true splice loss)
  - Acceptance values
- Reflectance
  - Number of wavelengths
  - Acceptance values
- ORL
- Dispersion

Standard Summary (14-37)

- Building Wiring Standard, TIA/EIA-568-C.0
- Building Wiring Standard, TIA/EIA-568-C.1
- Building Wiring Standard, TIA/EIA-568-C.3
  - Multimode Method B of ANSI/TIA/EIA-526-14-A
  - Singlemode Method A.1 of ANSI/TIA/EIA-526-7-A
- TIA/EIA-568-C.3
  - Multimode Category 1 source
  - Future: EF source or launch cable
- FOTP-107-A reflectance testing with reflectance test set
- FOTP-61 for attenuation rate measurement with OTDR

End Of Chapter

Questions?
Comments?
Observations?

Fiber Cable Testing, Certification, And Troubleshooting — Chapter 11: OTDR Testing

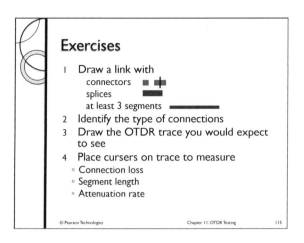

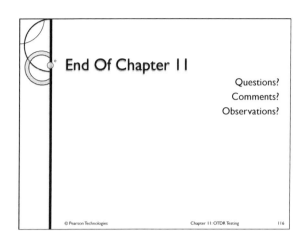

OTDR Maps For OTDR Testing, Chapter 11

Reel Sets A, B, C: all lengths ~1000 m

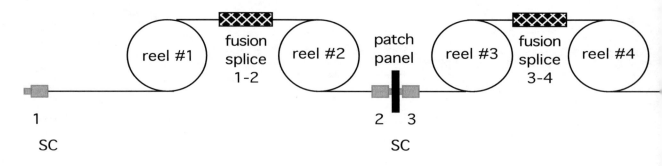

Reel Set D

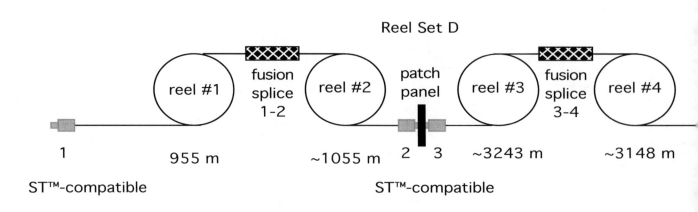

OTDR Measurement Data Form

LINK #	A					LINK #	B			
Direction	From reel 1		use √			Direction	From reel 1		use √	
	From reel 4		use √				From reel 4		use √	
Singlemode ?						Singlemode ?				
Multimode ?						Multimode ?				

Reel #	1	2	3	4		Reel #	1	2	3
Cursar left					m	Cursar left			
Cursar right					m	Cursar right			
Length					m	Length			

Attenuation rate					dB/km	Attenuation rate				
Cursar left					m	Cursar left				
Cursar right					m	Cursar right				

	1-2	3-4			1-2	3-4
Splice loss			dB	Splice loss		
Cursar left			m	Cursar left		
Cursar right			m	Cursar right		

	1	2-3	4			1	2-3	4
Connector loss				dB	Connector loss			
Cursar left				m	Cursar left			
Cursar right				m	Cursar right			

Type of splice	1-2	3-4			Type of splice	1-2	3-4	
Mechanical			use √		Mechanical			use √
Fusion			use √		Fusion			use √

Type of connector	2-3	1, 4			Type of conn	2-3	1, 4	
Radius			use √		Radius			use √
APC			use √		APC			use √

© Pearson Technologies Inc.   6/15/14

OTDR Measurement Data Form

LINK #	C					LINK #	D			
Direction	From reel 1		use √			Direction	From reel 1		use √	
	From reel 4		use √				From reel 4		use √	

Singlemode ?						Singlemode ?				
Multimode ?						Multimode ?				

Reel #	1	2	3	4		Reel #	1	2	3
Cursar left					m	Cursar left			
Cursar right					m	Cursar right			
Length					m	Length			

Attenuation rate					dB/km	Attenuation rate			
Cursar left					m	Cursar left			
Cursar right					m	Cursar right			

	1-2		3-4				1-2		3-4
Splice loss					dB	Splice loss			
Cursar left					m	Cursar left			
Cursar right					m	Cursar right			

	1	2-3	4				1	2-3	4
Connector loss					dB	Connector loss			
Cursar left					m	Cursar left			
Cursar right					m	Cursar right			

Type of splice	1-2	3-4			Type of splice	1-2	3-4	
Mechanical			use √		Mechanical			use √
Fusion			use √		Fusion			use √

Type of connector	2-3	1, 4			Type of conn	2-3	1, 4	
Radius			use √		Radius			use √
APC			use √		APC			use √

© Pearson Technologies Inc.

OTDR Measurement Data Form

LINK # A

Direction	From reel 1		use √
	From reel 4		use √

Singlemode ?	
Multimode ?	

Reel #	1	2	3	4	
Cursar left					m
Cursar right					m
Length					m

Attenuation rate					dB/km
Cursar left					m
Cursar right					m

	1-2	3-4	
Splice loss			dB
Cursar left			m
Cursar right			m

	1	2-3	4	
Connector loss				dB
Cursar left				m
Cursar right				m

Type of splice	1-2	3-4	
Mechanical			use √
Fusion			use √

Type of connector	2-3	1, 4	
Radius			use √
APC			use √

LINK # B

Direction	From reel 1		use √
	From reel 4		use √

Singlemode ?	
Multimode ?	

Reel #	1	2	3
Cursar left			
Cursar right			
Length			

Attenuation rate			
Cursar left			
Cursar right			

	1-2	3-4
Splice loss		
Cursar left		
Cursar right		

	1	2-3	4
Connector loss			
Cursar left			
Cursar right			

Type of splice	1-2	3-4	
Mechanical			use √
Fusion			use √

Type of conn	2-3	1, 4	
Radius			use √
APC			use √

© Pearson Technologies Inc.   6/15/14

OTDR Measurement Data Form

LINK #	C					LINK #	D			
Direction	From reel 1				use √	Direction	From reel 1			use √
	From reel 4				use √		From reel 4			use √
Singlemode ?						Singlemode ?				
Multimode ?						Multimode ?				

Reel #	1	2	3	4	
Cursar left					m
Cursar right					m
Length					m
Attenuation rate					dB/km
Cursar left					m
Cursar right					m

	1-2		3-4	
Splice loss				dB
Cursar left				m
Cursar right				m

	1	2-3	4	
Connector loss				dB
Cursar left				m
Cursar right				m

Type of splice	1-2	3-4	
Mechanical			use √
Fusion			use √

Type of connector	2-3	1, 4	
Radius			use √
APC			use √

Reel #	1	2
Cursar left		
Cursar right		
Length		
Attenuation rate		
Cursar left		
Cursar right		

	1-2	3-4
Splice loss		
Cursar left		
Cursar right		

	1	2-3	4
Connector loss			
Cursar left			
Cursar right			

Type of splice	1-2	3-4	
Mechanical			use √
Fusion			use √

Type of conn	2-3	1, 4	
Radius			use √
APC			use √

© Pearson Technologies Inc. 6/15/14

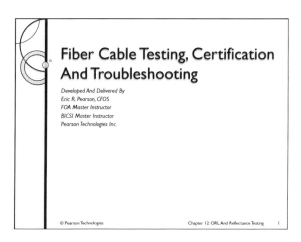

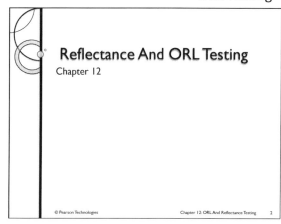

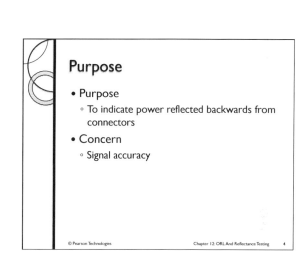

Purpose

- Purpose
  - To indicate power reflected backwards from connectors
- Concern
  - Signal accuracy

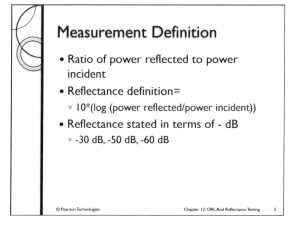

Measurement Definition

- Ratio of power reflected to power incident
- Reflectance definition=
  - 10*(log (power reflected/power incident))
- Reflectance stated in terms of - dB
  - -30 dB, -50 dB, -60 dB

Measurement Definition

- Less is better: -60 dB is better than -50 dB
- For reference: -60 dB means one millionth of the power coming to the connector is reflected backwards!
- Very small power levels
  - E.g., a -60 dB reflectance with 1 mW incident means 1 nanowatt of reflected power

Reflectance

- Performed on
  - Patch cords
  - Pigtails
- Two methods
  - FOTP-107 with a reflectance test set
  - FOTP-8- with an OTDR

Equipment

- A reflectance test set [FOTP-107]
- A reference lead
- Low loss, clean barrels
- A mandrel with a diameter appropriate to the wavelength of the test
- Connector cleaning supplies

Reflectance Test Set Structure

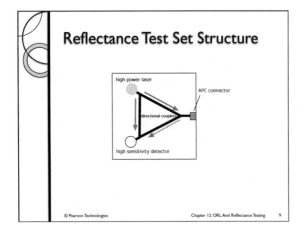

Procedure

- Clean all connectors
- Set a reference power level (done automatically at turn on)
- Null out power from internal reflections in the test set
- Check the reference lead against a known low reflectance reference lead
- Connect the reference lead to the connector to be tested
- If required, wrap the opposite end of the jumper around a mandrel

Verify Low Reflectance Of Test Lead

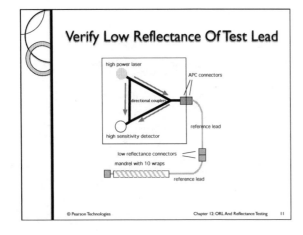

Test Reflectance

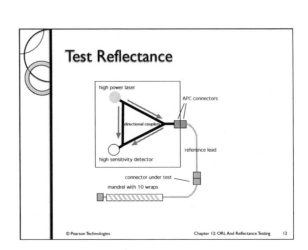

Fiber Cable Testing. Certification, And Troubleshooting — Chapter 12: Reflectance And ORL Testing

Cleanliness Critical
- We use Electro-Wash Px (Opinion)
- Others use 99 % isopropyl alcohol

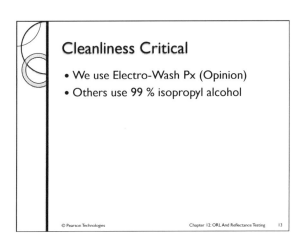

Acceptance Values
- Multimode connectors: ≤ -20db
- Singlemode connectors: ≤ -26 dB
- Analog CATV applications: ≤ -55 dB

  - From TIA/EIA-568-C

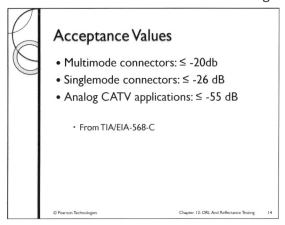

OPTICAL RETURN LOSS TESTING [ORL]

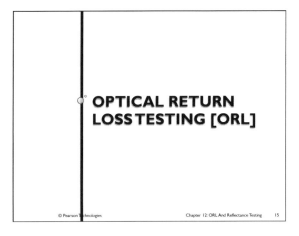

Optical Return Loss Test
- Performed to verify total power reflected and back scattered to transmitter is sufficiently low
- Performed in a manner similar to that of reflectance testing

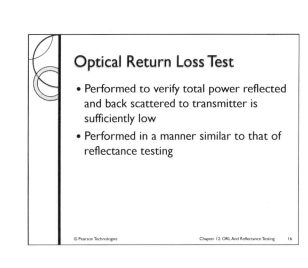

Step 1: Null Out Power

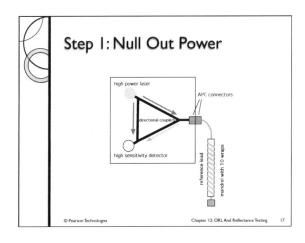

Step 2: Test ORL

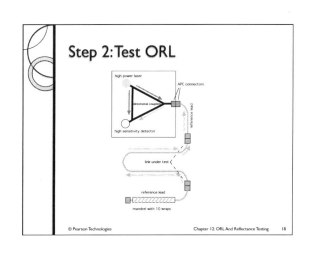

Fiber Cable Testing. Certification, And Troubleshooting — Chapter 12: Reflectance And ORL Testing

Cleanliness Critical
- We use Electro-Wash Px (Opinion)
- Others use 99 % isopropyl alcohol

Practicalities
- High ORL often due to dirty connectors at end near ORL test unit [opinion]
- Cleaning can reduce ORL to acceptable levels
- Acceptable level defined by transmitter manufacturer or by company policy
- ≤ -30 dB a typical value

Define Test Requirements
- Specify insertion loss
  - Method
  - Number of wavelengths
  - Number of directions (1 is acceptable)
  - Acceptance values
- OTDR
  - Number of wavelengths
  - Number of directions (2 necessary for true splice loss)
  - Acceptance values
- Reflectance
  - Number of wavelengths
  - Acceptance values
- ORL
- Dispersion

Standard Summary
- Building Wiring Standard, TIA/EIA-568-C.0
- Building Wiring Standard, TIA/EIA-568-C.1
- Building Wiring Standard, TIA/EIA-568-C.3
  - Multimode Method B of ANSI/TIA/EIA-526-14-A
  - Singlemode Method A.1 of ANSI/TIA/EIA-526-7-A
- TIA/EIA-568-C.3
  - Multimode Category 1 source
  - Future: EF source or launch cable
- FOTP-107-A reflectance testing with reflectance test set
- FOTP-61 for attenuation rate measurement with OTDR

End Of Chapter 12
Questions?
Comments?
Observations?

Reflectance And ORL Measurement Data Form

REFLECTANCE

UNICAM 1 Direction	1310 nm	1550 nm		UNICAM 2 Direction	1310 nm	1550 nm
Source at End A				Source at End A		
Source at End B				Source at End B		

ORL

LINK # A Direction	1310 nm	1550 nm		LINK # B Direction	1310 nm	1550 nm		LINK # C Direction	1310 nm	1550 nm		LINK # D Direction
Source at reel 1				Source at reel 1				Source at reel 1				Source at r...
Source at reel 4				Source at reel 4				Source at reel 4				Source at r...

Fiber Cable Testing, Certification And Troubleshooting

Developed And Delivered By
Eric R. Pearson, CFOS
FOA Master Instructor
BICSI Master Instructor
Pearson Technologies Inc.

Calculating Acceptance Values
Chapter 13

Interpretation Of Test Results

Introduction

- Verification of sufficiently low loss through the link
- Verification that each component in link has been properly and reliably installed

Four Parts

1. Obtain required data
2. Perform insertion loss calculations
3. Calculate insertion loss acceptance values
4. Calculate OTDR acceptance values

Required Information

- Accurate map
- Attenuation rate, maximum
- Attenuation rate, typical
- Connector loss, maximum
- Connector loss, typical
- Splice loss, maximum
- Splice loss, typical

Step 2: Perform Insertion Loss Calculation

- Calculation depends on the Method used (Ch. 10)
- Use Method B
  ◦ Treats two end connectors as two pairs

Fiber Cable Testing, Certification, And Troubleshooting
Chapter 13: Calculating Acceptance Values

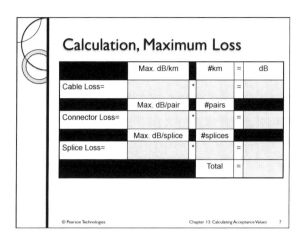

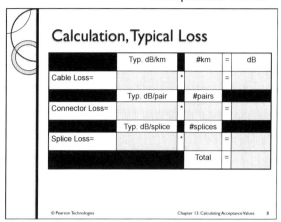

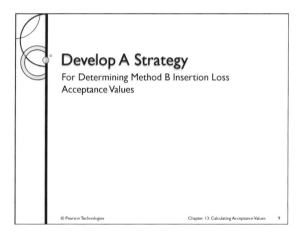

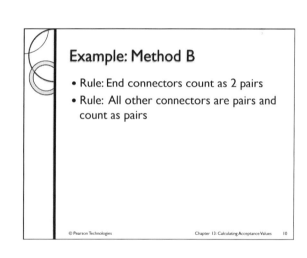

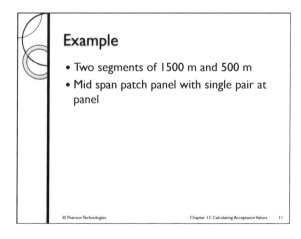

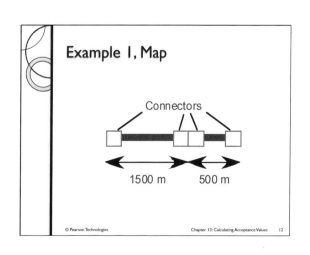

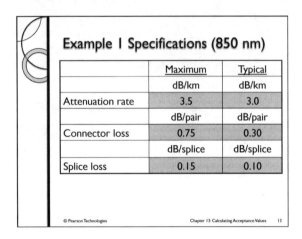

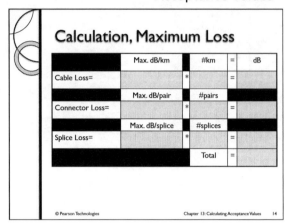

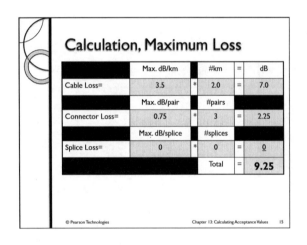

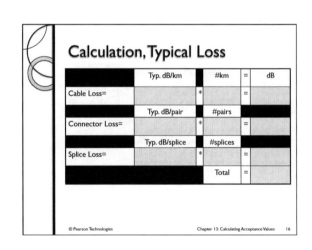

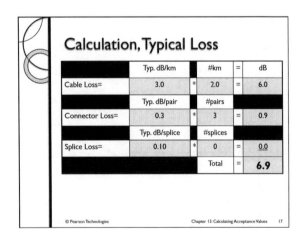

What Happens If You Accept Maximum Loss?
- You allow up to 2.35 dB of excess loss.
- Excess loss occurs only through installation errors
- Installation errors reduce network reliability

Fiber Cable Testing, Certification, And Troubleshooting
Chapter 13: Calculating Acceptance Values

Conclusion

- Do not want to use maximum loss as an acceptance value
  - Risk: reduced reliability

Alternative

- Use typical loss?

Typical Loss

- There is a normal variation of loss around the typical value
- Consider use of the typical value as a maximum acceptance value
- This use results in rejection of properly installed connectors and links that are slightly above typical values
- Benefit: none
- Disadvantage: increased cost

Conclusion

- Do not want to use typical loss as a maximum acceptance value
  - Risk: increased cost

How To Calculate Insertion Loss Acceptance Value
For High Reliability

Insertion Loss Acceptance Value

- Is the highest insertion loss value that you will accept
- With a goal of high reliability
- Expect properly installed links to be closer to typical loss than maximum loss

Insertion Loss Acceptance Value

- Therefore, choose an acceptance value halfway between typical and maximum calculated values
- Call this 'mid point acceptance value'

Five Part Strategy

1. Expect measured loss to be closer to calculated typical value than to calculated maximum value
2. Calculate mid point acceptance value
3. Accept Test Values ≤ mid point acceptance value
4. Investigate Test Values ≥ mid point acceptance value and ≤ calculated maximum value
   - Probably Reject
5. Reject Test Values > calculated maximum value

Investigate Means

- Test with OTDR
- Inspect connectors with microscope
- Inspect connectors and splices with VFL

The Acceptance Value Is..

- Halfway Between Calculated Typical and Calculated Maximum Values
- = 1/2 * (6.9+9.25) = 8.075 dB

Mid-Point Acceptance Value

- Nothing magic
- This value allows minor mistakes

Example

- Multimode, 3M hot melt connectors are rated at typical loss of 0.3 dB /pair
- With over 10,000 installed in training and field work, this product never exceeds 0.4 dB /pair unless damage or dirt is visible on connector surface
- Mid point strategy allows connector loss up to 0.525 dB /pair
  - i.e., minor mistakes will be accepted

Fiber Cable Testing, Certification, And Troubleshooting
Chapter 13: Calculating Acceptance Values

Calculate OTDR Acceptance Values

- Use same midpoint strategy to calculate acceptance values for connector losses, splice losses and attenuation rates
  - Accept test values less than or equal to the midpoint acceptance value
  - Investigate test values greater than the midpoint acceptance value and less than the maximum value
  - Reject test values greater than the maximum value

OTDR Acceptance Values

- Attenuation rate acceptance value= (maximum rate + typical rate)/2
- Connector pair acceptance value= (maximum loss + typical loss)/2
- Splice loss acceptance value= (maximum loss + typical loss)/2

Example 1 Specifications (850 nm)

	Maximum	Typical
	dB/km	dB/km
Attenuation rate	3.5	3.0
	dB/pair	dB/pair
Connector loss	0.75	0.30
	dB/splice	dB/splice
Splice loss	0.15	0.10

Examples

- Attenuation rate acceptance value= (3.5 + 3.0)/2= 3.25 dB/km
- Connector pair acceptance value= (0.75 + 0.30)/2= 0.525 dB/pair
- Splice loss acceptance value=
- (0.15 + 0.10)/2= 0.125 dB/splice

Additional Trace Requirement

- Every cable segment must exhibit a uniform loss (i.e., be a straight line)
- A properly designed, properly manufactured, properly installed cable segment always exhibits a straight-line trace
- Any deviation from a straight trace indicates an installation error at location of deviation

Alternative Strategy
For Maximum Reliability

Three Steps

- Comparison of OTDR attenuation rates of each segment prior to installation to those rates after installation
  - Allow no increase in attenuation rate
- Visual inspection of all connectors with acceptance requirements for the core, cladding, and ferrule surface
  - Require round, clear, featureless, and flush cores
  - Require clean cladding and ferrule surface
- Use of a maximum connector loss value less than the mid point value
  - Example: 3M multimode connector at ≤0.4 dB /pair

Disadvantage

- High cost

Summary
MIDPOINT ACCEPTANCE VALUES

Acceptance Values

- Insertion loss
  - Insertion loss= (maximum loss + typical loss)/2
- OTDR
  - Attenuation rate= (maximum rate + typical rate)/2 plus uniform segment traces
  - Each segment has uniform attenuation (straight line trace)
  - Connector loss= (maximum loss/pair + typical loss/pair)/2
  - Splice loss= (maximum loss/splice + typical loss/splice)/2

Exercises 1 and 2, Section 15.8

- Determine insertion loss values
  - Maximum
  - Minimum
- Determine acceptance values
  - Insertion loss
  - OTDR connector loss
  - OTDR attenuation rate
  - OTDR splice loss

Cell Tower Exercises, Section 9

- For all 3 towers, determine insertion loss values
  - Maximum
  - Typical
- For all 3 towers, determine acceptance values
  - Insertion loss
  - OTDR connector loss
  - OTDR attenuation rate
  - OTDR splice loss

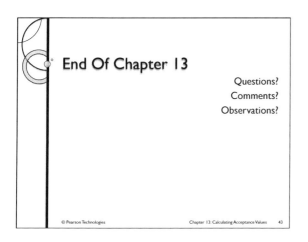

Made in the USA
Monee, IL
18 August 2025